Robots and Communication

Other Palgrave Pivot titles

Hyunjung Lee: Performing the Nation in Global Korea: Transnational Theatre

Creso Sá M. and Andrew J. Kretz: The Entrepreneurialism Movement and the University

Emma Bell: Soft Power and Freedom under the Coalition: State-Corporate Power and the Threat to Democracy

Ben Ross Schneider: Designing Industrial Policy in Latin America: Business-State Relations and the New Developmentalism

Tamer Thabet: Video Game Narrative and Criticism: Playing the Story

Raphael Sassower: Compromising the Ideals of Science

David A. Savage and Benno Torgler: The Times They Are A Changin': The Effect of Institutional Change on Cooperative Behaviour at 26,000 ft over Sixty Years

Mike Finn (editor): The Gove Legacy: Education in Britain after the Coalition

Clive D. Field: Britain's Last Religious Revival? Quantifying Belonging, Behaving, and Believing in the Long 1950s

Richard Rose and Caryn Peiffer: Paying Bribes for Public Services: A Global Guide to Grass-Roots Corruption

Altug Yalcintas: Creativity and Humour in Occupy Movements: Intellectual Disobedience in Turkey and Beyond

Joanna Black, Juan Carlos Castro, and Ching-Chiu Lin: Youth Practices in Digital Arts and New Media: Learning in Formal and Informal Settings

Wouter Peeters, Andries De Smet, Lisa Diependaele and Sigrid Sterckx: Climate Change and Individual Responsibility: Agency, Moral Disengagement and the Motivational Gap

Mark Stelzner: Economic Inequality and Policy Control in the United States

Michelle Bayefsky and Bruce Jennings: Regulating Preimplantation Genetic Diagnosis in the United States

Eileen Piggot-Irvine: Goal Pursuit in Education Using Focused Action Research

Serenella Massidda: Audiovisual Translation in the Digital Age: The Italian Fansubbing Phenomenon

John Board, Alfonso Dufour, Yusuf Hartavi, Charles Sutcliffe and Stephen Wells: Risk and Trading on London's Alternative Investment Market: The Stock Market for Smaller and Growing Companies

Franklin G. Mixon, Jr: Public Choice Economics and the Salem Witchcraft Hysteria

Elisa Menicucci: Fair Value Accounting: Key Issues Arising from the Financial Crisis

DOI: 10.1057/9781137468376.0001

palgrave▸pivot

Robots and Communication

Eleanor Sandry

palgrave macmillan

© Eleanor Sandry 2015

All rights reserved. No reproduction, copy or transmission of this publication may be made without written permission.

No portion of this publication may be reproduced, copied or transmitted save with written permission or in accordance with the provisions of the Copyright, Designs and Patents Act 1988, or under the terms of any licence permitting limited copying issued by the Copyright Licensing Agency, Saffron House, 6–10 Kirby Street, London EC1N 8TS.

Any person who does any unauthorized act in relation to this publication may be liable to criminal prosecution and civil claims for damages.

The author has asserted her right to be identified as the author of this work in accordance with the Copyright, Designs and Patents Act 1988.

First published 2015 by
PALGRAVE MACMILLAN

Palgrave Macmillan in the UK is an imprint of Macmillan Publishers Limited, registered in England, company number 785998, of Houndmills, Basingstoke, Hampshire RG21 6XS.

Palgrave Macmillan in the US is a division of St Martin's Press LLC, 175 Fifth Avenue, New York, NY 10010.

Palgrave Macmillan is the global academic imprint of the above companies and has companies and representatives throughout the world.

Palgrave® and Macmillan® are registered trademarks in the United States, the United Kingdom, Europe and other countries.

ISBN: 978–1–137–46838–3 EPUB
ISBN: 978–1–137–46837–6 PDF
ISBN: 978–1–137–46836–9 Hardback

A catalogue record for this book is available from the British Library.

A catalog record for this book is available from the Library of Congress.

www.palgrave.com/pivot

DOI: 10.1057/9781137468376

To Dave (a nonrobotic human) and Jemma (a nonrobotic dog)

Contents

List of Figures	ix
Acknowledgements	x
Introduction	1
What is a robot?	3
What is communication?	4
Structure and scope	8

Part I   From Data to Dogs

1 Designing Robots to Communicate with Humans	12
Science fiction origins	13
Humanoid robots in real life	14
Emotional robots	16
Communicating with Data	18
Avoiding the 'uncanny valley'	22
Communicating with Kismet	23
Critiquing the pursuit of sameness	28
2 Human-Animal Communication	31
Animal communication in fiction	32
Communicating with Clever Hans	34
Recognising individual animals	36
Communicating about relationships	39
Companion species and communicating about what's 'out there'	39

Learning to value anthropomorphism	41
From human-animal communication to understanding robots	43

Part II Communicating with Non-Humanoid Robots

3 Encountering Otherness	**46**
The Autonomous Light Air Vessels	47
Levinasian encounters	50
Formalising the idea of communication as more than language	54
Making space for others and their differences	55
The ALAVs in version 2.0	58
4 Stories and Dances	**62**
The Fish-Bird project	63
Interruptions in being and in saying	64
Dialogues and dynamic systems	68
Employing both discrete state and dynamic systems models	70
5 Collaboration and Trust	**74**
AUR the robotic desk lamp	76
Collaborating with AUR	77
Acting theory as an alternative perspective	80
Expression and 'face'	83
Trust, respect and interruption in dynamic interactions	85

Part III Rethinking Robots and Communication

6 Humans, Animals and Machines	**89**
The value of blurred yet meaningful boundaries	90
Recognising the difference of humanoid robots	92
Non-humanoid robots and difference	96
Activity and agency within a particular context	97
7 Communication, Individuals and Systems	**100**
Rehearsal or training and fluency	101
Reconsidering Kismet as part of an interaction system	102

Long-term interactions with nonautonomous robots   103
Moving closer to home with robotic floor cleaners   107
Asymmetry, responsibility and reciprocity   110

Conclusion   113

Bibliography   116

Index   125

List of Figures

1.1	Kismet	24
3.1	The ALAVs (version 1.0)	48
4.1	The Fish-Bird project	64
5.1	AUR	76
7.1	Packbot with soldier	105

Acknowledgements

Writing this book would not have been possible without the help of many people. For the first few years my PhD supervisors, Chantal Bourgault du Coudray and Tanya Dalziell, were the people I relied upon the most for their immensely valuable suggestions and encouragement. The examiners of my thesis also provided helpful comments and ideas for further consideration. More recently it has been my colleagues at Curtin University who inspired many of the additional thoughts and writing that have been added in the past year or so. The feedback from the anonymous reviewer of my book proposal was also invaluable.

A great deal of the research represented within this book was funded by an Australian Postgraduate Award, in association with a top-up scholarship provided by the University of Western Australia. In 2013, I benefited from a six-month CCI Research Fellowship in the Centre for Culture and Technology at Curtin, working with John Hartley, and more recently the Department of Internet Studies and the School of Media, Culture and Creative Arts have done all that they could to ensure I had the time to complete this book.

Finally, and by no means least, I would like to thank my family and friends who have had to put up with me, and with the robots, for a very long time. Although I have dedicated this book to the ones who have had to share living space with me throughout the writing process, I would also like to give special thanks to my mum, who lives at a distance. She not only read the whole of my PhD thesis in its final draft to look for errors and nonsensical passages (thus feeding her comments into this book as well), but has also been happy to discuss ideas about robots with me ever since.

palgrave▸pivot

www.palgrave.com/pivot

Introduction

Abstract: *In order to prepare the reader for an analysis of human interactions with a wide variety of forms of robot, the introduction first explores what constitutes a robot. It then goes on to outline the different traditions of communication theory that are employed in the book's analysis. Finally, it explains the structure and scope of the book.*

Sandry, Eleanor. *Robots and Communication*. Basingstoke: Palgrave Macmillan, 2015. DOI: 10.1057/9781137468376.0004.

This book considers human-robot interactions as models of communication with the aim of re-evaluating the presence of otherness in communicative encounters. Its focus is therefore on finding ways to *value* the differences between communicators as opposed to regarding them as *problems* that must be overcome. This re-evaluation is difficult to achieve because, from its etymological roots to its metaphorical articulation in everyday language, communication is often linked with the accurate transmission of information and/or the construction and re-construction of shared social understandings (Reddy, 1979; Lakoff, 1980; Chang, 1996; Peters, 1999). From these perspectives, successful communication is positioned as a bridge between self and other, founded on their commonalities and seeking to develop those commonalities further.

In terms of reconsidering the value of difference, robots are intriguing communicators because they appear in such a variety of forms. Robots are sometimes created to be as humanlike as possible, but on other occasions are 'overtly other' in design. Within the continuum between these two extremes some robots are animal-like, while others resemble everyday objects that move and respond to people and their surroundings in unexpected ways. Therefore, at one end of the spectrum human interactions with humanoid robots illustrate the effects of accepting that commonality is key to effective communication, as well as exposing some of the limitations of this perspective. Towards the 'other' end of the spectrum, human interactions with robots that are not humanlike (or even animal-like) demonstrate the possibilities of communication that values difference, while nonetheless supporting effective collaborations between humans and machines. It is difficult to know how best to refer to the 'overtly other' end of the spectrum of design as a group, since broadly these robots cannot be defined simply as machinelike or object-like; they might not resemble anything familiar at all. Represented in this book are blimp-like, wheelchair-like, lamp-like and tank-like robots, along with one that is shaped like a shallow box (the iRobot Braava). For the purposes of this book I have therefore decided to refer to this disparate group of robots as 'non-humanoid', as a shorthand for 'not overtly humanlike or animal-like', because much of the book focuses on considering the possibilities of communication between humans and nonhuman others.

The decision to examine communication between humans and many different types of robot, from humanoid to non-humanoid, was inspired by John Durham Peters' suggestion that '[b]y exploring our strangest partners' it is possible 'to illuminate the strangeness that occurs in the most familiar settings' (1999, p. 231). An exploration of the interactions between humans and robots, and the interplay of familiarity and strangeness within those interactions, supports this book's argument that difference can be of value in communication, and collaboration, with all kinds of others.

Before moving on to provide an overview of the book's parts and chapters, it is worth discussing briefly the origins and varied present-day uses of the term 'robot', as well as offering an overview of the categories of communication theory that are employed in its analyses of human-robot interactions.

What is a robot?

The word 'robot' was first used in 1920 in Karel Capek's play *R. U. R.* (*Rossum's Universal Robots*), although Karel credited his brother Joseph as the originator of the term (Capek, 1933). The robots in *R. U. R.* are not metal machines, the construction that typifies many recent robots in fiction, as well as those in real life; instead, Rossum's robots are 'artificial people', constructed from organic components (Capek, 1920/2006, p. 7). While their internal structure is far simpler than that of a human, externally they look very much like the human factory workers they have been designed to replace (Capek, 1920/2006, p. 9). The term robot has, therefore, been associated with humanlike form from the very beginning, and the figure of the robot, which became more closely associated with human-shaped machines as opposed to organic artificial people, has become an icon for many science fiction writers, screen writers and film-makers.

In real-world contexts, the word robot is now used to describe a huge range of different forms of machine, some of which are radio-controlled, while others are partially or completely autonomous in their movements and actions. In fact, the use of the term has become so broad that it is difficult to define exactly what is meant when a machine is described as a robot. It is also very difficult to clarify, in any universally accepted way, what attributes or abilities differentiate a robot from any other machine.

In general, a robot might be best regarded as a machine that appears to have some level of agency, and therefore seems to sense and respond to its surroundings. This is the case for all of the robots discussed in this book. However, given that this is a book about communication and robots, it should not be surprising that my particular interest is in robots that people regard as communicating with them in some way. Contrary to what might be assumed, a robot does not need to have a high level of autonomy to be regarded as a communicative partner. This is made particularly clear in Chapter 7, during the discussion of the relations between soldiers and Explosive Ordnance Disposal (EOD) robots that are currently almost always under the direct control of a human operator.

Stories about robots and developments in robotics are now reported on a regular basis, not only in technology-focused news media, but also in mainstream television and news publications. Even having a robot in the home is no longer just a technological fantasy for some people, although robotic vacuum cleaners are still a long way from the humanoid robotic helpers described in some science fiction. Robots are becoming more visible in workplaces as well and are sent into dangerous situations such as war, rescue or exploration. The use of robotic technologies is also increasingly directed towards providing care robots in hospitals and for assisted home living. In addition, robots have been introduced into educational environments and are encountered in public spaces, where they may be part of interactive art installations, or may act as informative museum and exhibition guides. As robots become more commonplace, the question of how people and robots can communicate with one another becomes increasingly important, whether robots simply encounter people in shared spaces, are required to work with them in teams or are positioned as their rescuers, carers or companions.

In order to explore the possibilities of human interactions with robots – from familiar to radically other – this book employs a range of communication theories to offer different perspectives on what happens when humans and robots meet and communicate, whether in scientific laboratories, art installations or science fiction. This exploration of human-robot interactions also circles back and provides new ways to think about communication in theory and in practice, offering some useful ways to rethink the presence of otherness in communicative processes and systems.

What is communication?

In discussing communication, this book draws upon the seven traditions of communication theory identified by Robert T. Craig (1999): rhetorical, concerned with all forms of discourse; semiotic, studying the use of signs for intersubjective mediation; phenomenological, acknowledging the experience of otherness in authentic relationships; cybernetic, describing communication in terms of information processing models; sociocultural, analysing social and cultural contexts in the production and reproduction of social order; sociopsychological, concentrating on the context of subjective reactions to expression, interaction and influence; and critical, concerning the development of reasonable and rational discourse.

These traditions of communication 'offer distinct ways of conceptualizing and discussing communication problems and practices' and therefore provide different lenses through which to analyse communicative events (Craig, 1999, p. 120). In this book I follow Craig's suggestion that it is valuable to explore the 'dialogue' between these theories in an attempt to more 'fully engage with the ongoing practical discourse (or metadiscourse) about communication in society' with a particular focus on questions about otherness and difference (1999, p. 120). Of course, a book of this length cannot hope to provide an exhaustive analysis of the human-robot interactions it discusses from all seven points of view, but all traditions play their part in what follows, with a particular focus on the interplay between the cybernetic, semiotic, sociocultural and phenomenological traditions. In addition, since the rhetorical tradition has 'shifted from a focus on oratory' (and therefore the use of spoken language) to cover the use of 'every kind of symbol' it would seem to have a part to play in all communication, and is not overtly mentioned here (Littlejohn and Foss, 2011, p. 64).

Whether communication is understood in terms of rhetorical force, the accurate transmission or exchange of information, gaining influence through persuasion, rational argument towards agreement or the adoption of shared social understandings, it is most often judged to rely upon and to foster similarities between communicators. However, thinking about communication in this way reduces one's awareness of the potential of the other's point of view and devalues the importance of understanding and knowledge about the world that is different from one's own. As a number of communication scholars have argued, taking this stance against diversity can be seen as violent to the other, forcing them to conform to what is judged as the norm, as opposed to being open to their otherness (Peters, 1999; Pinchevski, 2005). In terms of robot design, these understandings of communication form part of the reasoning behind the drive to create humanoid robots, because it is assumed that human-robot interactions will be easier and will work better when robots are designed to be familiar, humanlike others.

The careful recognition of otherness is most clearly valued by the phenomenological tradition of communication theory (Peters, 1999; Pinchevski, 2005). Theories within this tradition are focused upon the development of ethical communication with the other, and therefore with the view that difference is an integral part of communicative processes, as opposed to a difficulty that must be overcome in order for successful communication to occur. However, the way that this tradition focuses on being open to otherness, while acknowledging the impossibility of ever completely comprehending the other, makes it very difficult to consider in relation to other theoretical traditions in the field. As Craig notes, '[p]henomenology, from a rhetorical point of view, can seem hopelessly naïve or unhelpfully idealistic in approaching the practical dilemmas that real communicators must face' (1999, pp. 139–140). Extending this observation, it seems reasonable to suggest that an appraisal of the phenomenological tradition as somewhat naive and idealistic might be shared more generally from the perspective of all other communication traditions that focus upon success, whether in terms of passing information accurately, influencing others or creating social cohesion.

It is also noticeable that scholars who write about the phenomenological tradition tend to use emotive examples to illustrate the call of the other from a position of need. This strategy emphasises that the self is in a position of responsibility, having a choice over whether to take an ethical stance towards the other (Levinas, 1969, p. 251; Lingis, 1994, p. 12; Smith, 1997, p. 332). These discussions of communication are therefore distanced from the more usual everyday concerns of communicators. In contrast, although they are easily placed as others, the robots discussed in this book are not emotive in the same sense as a sick child, for example. In addition, although these robots in themselves are not everyday objects, examples of their communication with humans are clearly positioned within this space. Thus, the use of robot illustrations offers various ways to draw out phenomenological concerns, in particular relating to responsibility and respect, within contexts describing communication that might take place as part of everyday life. An analysis of the phenomenological tradition through human-robot examples therefore opens up ways to value phenomenological ideas alongside more familiar, or traditional, conceptions of communication in those situations.

Craig also argues that 'from a phenomenological point of view', the rhetorical tradition, and given the contention that it underlies other theories I would also suggest many other traditions, 'can seem unduly cynical or pessimistic about the potential for authentic human contact' (1999, p. 140). Amit Pinchevski's work in particular can be associated with such a phenomenological perspective, since it suggests not only that traditional communication theory suffers from a 'conceptual blindness' to otherness, but also that, as a result, it is overtly violent to the other (2005, pp. 29–62). Craig notes that '[w]hen rhetoric and phenomenology are combined, the result is typically an antirhetorical rhetoric in which persuasion and strategic action are replaced by dialogue and openness to the other'; or, sometimes 'a hermeneutical rhetoric in which the roles of theory and method in communicative practice are downplayed' (1999, p. 140). In this book, I do not generally attempt to combine theories, but rather analyse human-robot interactions from a number of perspectives. This approach helps to draw out the broader possibilities of each communicative situation, by considering communicators, their relationships and the communication systems of which they are a part, in ways that preserve otherness and difference, even while acknowledging the partial connection that develops between human and robot.

It is the non-humanoid robots discussed in Part II, with their overtly recognisable differences from humans, that are particularly helpful in moving phenomenological ideas away from contexts stressing human need, and towards a more general sense of the importance of otherness in encounters. In particular, robots that are created as part of art installations can be understood to push the boundary of what constitutes communication. Such robots are designed in part to entertain visitors, but they also raise questions relating to the possibilities of encounters with very different and unexpected others. There is less of a sense that such machines must communicate clearly and directly with their visitors, combined with more openness to the expectation that different people might interpret the robots in very different ways from one another. In the context of art installations there is therefore an understanding that the wide range of reactions non-humanoid robots might provoke is desirable.

The perspective on communication in this book regards both verbal and nonverbal communications as being equally worthy of attention. In addition, while viewing communication in terms of discrete moments or static processes can be valuable, considering dynamic systems of communication offers new ways to explore what happens during interactions. Regarding communication as a dynamic process of overlapping interchange is particularly relevant when thinking about nonverbal communication channels and the way that partial understandings develop between communicators (as opposed to the level of complete comprehension favoured by traditional theory with its goal of increasing what communicators have in common). A sense of partial understanding is supported by the recognition that aspects of a non-humanoid other can be read by comparison with past experiences of human or animal interactions, while also acknowledging the importance of the alterity of the machine. Partial understanding can therefore be linked with the ideas of tempered anthropomorphism and zoomorphism set out in the later chapters of this book. Thinking about dynamic systems of communication also leads to the consideration of communication over time, the importance of history and backstory as well as the value of shared experience gained when communicators train, learn and work together.

Structure and scope

This book is divided into three parts, the first of which contains two chapters and provides an overview of how communication is framed in relation to humanoid robots, as well as the possibilities of communication between humans and animals. In Chapter 1, the pursuit of humanlike form is analysed in both fictional and real-life contexts. Amongst other justifications, people committed to building humanoid robots argue that these robots are best suited to work in human environments and to communicate with humans. Two paths in humanoid robot design are considered, but both involve understandings of communication that value commonality over difference. This chapter draws on critiques of the pursuit of sameness found in the work of communication scholars, which is not much discussed in robotics, to destabilise the assumption that humanlike form is the best form for communicative robots.

Chapter 2 pauses the book's consideration of robots to explore human-animal communication, since nonhuman animals are an important part of many people's lives, acting for some not only as companions but also as co-workers. Human-animal communication is described in various, often idealised, ways in fiction, but in real-life situations an analysis of human-dog communication demonstrates the importance of attending to the smallest of nonverbal signs over periods of dynamic communication. This chapter highlights the possibilities for humans and animals, in particular dogs, to work together in teams, employing their specific skills to allow the team to perform tasks that neither human nor dog could complete alone. The ideas about communication developed in this chapter, which demonstrate how communication does occur in the presence of overt otherness, support the possibilities of human interactions with non-humanoid robots. A brief appraisal of the design and development of animal-like robots is included at the end of the chapter.

Part II contains three chapters, which provide a progressive exploration of human interactions with non-humanoid robots from initial meetings to prolonged engagement in a working team. Chapter 3 concentrates on theorising the encounter between human and robot, identifying moments when communication occurs, often using nonverbal communication channels at least initially. The chapter discusses two versions of the Autonomous Light Air Vessels (ALAVs) art installation, within which blimp-like robots interact with one another and with visitors. This chapter extends Levinas' conception of self-other encounters to consider nonhumans, including robots, in order to emphasise how communication can be understood to draw the self and the other into proximity while retaining the differences between them.

In Chapter 4, the discussion moves beyond the initial encounter, to consider how dynamic interactions support communication with robots where those communications are also framed by a backstory. In this chapter the focus is on how interactions can be understood in terms of both dialogue and overlapping continuous systems of interchange. Levinas' theory is further extended in this chapter, to highlight the interruptions in being and in saying that occur during interactions with Fish and Bird, the wheelchair-like robots discussed throughout the chapter.

Finally, Chapter 5 considers what happens when humans and robots learn to complete tasks together as a team. In particular, the chapter considers human interactions with AUR, the robotic desk lamp. In this example, elements of verbal and nonverbal communication are combined in a dynamic communication that involves paying attention to each other as well as to the task at hand. The chapter considers communication with AUR in terms of the companion species relation proposed by Donna Haraway, which was discussed in relation to human-dog agility teams in Chapter 2.

The two chapters in Part III draw together some of the overarching ideas of the book. Chapter 6 considers the implications of human communication with nonhuman others for the categories human, animal and machine. This chapter argues that, while the boundaries between these types of being are becoming increasingly blurred, they are nevertheless still meaningful. The chapter goes on to consider ways of assigning agency to nonhuman others on the basis of their activity in situations, while also recognising the difference between human activities and nonhuman activities.

Chapter 7 concentrates on exploring some ideas about the relationship between individuals and systems in thinking about communication. The chapter discusses long-term interactions with robots outside of laboratories and art installations, identifying the value of respect and trust in collaborative partnerships with robots. This is developed into a consideration of how responsibility is shared across collaborative teams, even when the team members are in an asymmetrical relationship.

Finally, the short conclusion to this book explains the basis for its somewhat eclectic analysis, which uses a range of traditions of communication theory, as well as considering the overarching conceptions of discrete state and dynamic systems methodologies.

Part I
From Data to Dogs

1
Designing Robots to Communicate with Humans

Abstract: *In Chapter 1, the pursuit of humanlike form is analysed in both fictional and real-life contexts. Amongst other justifications, people committed to building humanoid robots argue that these robots are best suited to work in human environments and to communicate with humans. Two paths in humanoid robot design are considered, but both involve understandings of communication that value commonality over difference. This chapter draws on the critiques of the pursuit of sameness found in the work of communication scholars, which is not much discussed in robotics, to destabilise the assumption that humanlike form is the best form for communicative robots.*

Sandry, Eleanor. *Robots and Communication*. Basingstoke: Palgrave Macmillan, 2015. DOI: 10.1057/9781137468376.0006.

When designing robots to communicate with humans, roboticists draw on a number of different arguments for pursuing a level of human likeness in their designs. Some of these arguments are based on assumptions relating to the practical issues of making a machine that can operate in human-tailored environments. However, for some robots a further concern is to make the robot communicate in familiar humanlike ways. Achieving this design goal is not easy, but the goal itself is strongly reinforced by the widely accepted understanding that communication is based on what communicators have in common, with its success being related to increasing that commonality further. Though not discussed much in the field of robotics, communication theorists have critiqued the pursuit of commonality, and their arguments are explored in this chapter as a way to question the idea that humanlike form is the best path to adopt when creating communicative robots that might also be able to collaborate and work with people.

Science fiction origins

As was discussed in the Introduction, the word robot was used for the first time in Capek's play *R. U. R.* to refer to artificially created entities made from organic components and described as almost indistinguishable from humans. When staging the play, the robots are played by human actors who follow Capek's stage directions by making the robots appear 'slightly mechanical in their speech and movements, blank of expression, fixed in their gaze' (1920/2006, p. 5). The portrayal of the robot characters emphasises that they have no emotions or feelings, not even sensing pain in the early stages of the play's narrative. In spite of their rather staid behaviour, Rossum's robots still retain their humanlike abilities to act upon and respond to spoken orders and gestures, as well as being able to read written instructions. The original robots are therefore presented as the ultimate workers, easily instructed and able to perform all manual tasks that are asked of them without ever complaining.

The humanlike form of Rossum's robots not only explains the ease with which they can replace human workers in human-tailored environments within the play, but also supports Capek's use of them as figures that highlight the social and cultural divisions between factory workers and the privileged ruling class in his own society (Philmus, 2005, p. 103). The name robot, derived from the word 'robota' meaning 'forced labour' in Czech, positions these emotionless artificial beings as the servants or slaves of humankind and emphasises Capek's message that 'the industrial system treated human labourers as though they were machines' (Disch, 1998, pp. 8–9).

After this early introduction, the robot went on to become a familiar icon in science fiction, although more often made of metal than organic material. Narratives continued to concern robots as workers and perpetuated the assumption that humanlike form would enable robots to carry out what were once human tasks effectively, as well as supporting easy communication with people. This is notably the case in the fictional work and speculative essays of Isaac Asimov, a scientist and writer whose conception of robots and robotics remains influential today. Although Asimov did write a number of stories about non-humanoid robots, an example being the robotic cars in the short story 'Sally' (1953/1995, pp. 19–40), he was fascinated by the idea of making machines in human form. In his nonfiction essay, 'The Friends We Make', Asimov explains that his narratives are concerned largely with the human desire for a servant of a similar level of intelligence, a worker who is tireless, strong, contented and never bored (1977/1990).

In spite of the differences in their construction, there are clear similarities between Asimov's mechanical robot characters and the fictional Rossum's artificial humans. In Asimov's opinion, as in Rossum's, humanlike form and an average-sized body is an essential attribute to enable a robot to operate as an effective worker in an environment already tailored to humans (Asimov, 1977/1990, pp. 417–419; Capek, 1920/2006, p. 12). It is worth remembering that Capek's play dates from 1920, and Asimov originally wrote his essay in 1977, which may explain why the idea of making human environments and workplaces more accessible to all human individuals including, for example, those people using wheelchairs, was not considered significant by either author. In contrast, present-day requirements to address accessibility concerns for humans also allow wheeled or tracked robots to access those spaces more easily.

Humanoid robots in real life

In spite of the increased accessibility of present-day workplaces, Asimov's conception of robots as humanoid machines constructed from inorganic materials, and the idea that their humanlike form lends itself to working in human environments, is mirrored in the creation of real-life robots and discussions about their designs. For example, the design and size of Honda's bipedal walking machine, ASIMO, is understood 'to allow it to operate freely in the human living space', so that it can 'operate light switches and door knobs, and work at tables and work benches', while also making it 'people-friendly' (Honda, 2007, p. 15). The development of ASIMO has taken over a decade, and most of that time has been spent trying to perfect its walking, running and stair-climbing abilities. Publicity information regarding ASIMO suggests that its design enables it to work freely in unmodified human workplaces and homes. However, its ability to operate in constantly changing real-life environments is still fragile, and while Honda's websites contain videos of ASIMO's successes, a number of amateur videos on the Internet record some of the moments when ASIMO has experienced difficulties, including, for example, falling on stairs.

Towards the end of his essay, Asimov says that if people are to accept 'thinking partners – or, at the least, thinking servants – in the form of machines' it is likely that they 'will be more comfortable with them' and 'will relate to them more easily, if they are shaped like humans' (1977/1990, p. 419). He extends this concept even further, as he considers the possibility of humans living more closely with robots, suggesting that '[i]t will be easier to be friends with human-shaped robots than with specialized machines of unrecognizable shape' (Asimov, 1977/1990, p. 419). This appraisal is supported by the way that ASIMO's designers express their desire to make a robot that is people-friendly. It is clear that both Asimov and present-day designers of robots base the drive to make humanoid robots on factors that relate more closely to concerns of sociability, and the need to make robots more personable, than simply enabling them to work in human physical environments. In the course of this development, a rather uncomfortable juxtaposition between the idea of 'robots as servants' and 'robots as friends' arises, exacerbated by the origin of the term robot with Capek's concerns regarding factory workers and the idea of forced labour. These incompatible perspectives are still present in many discussions of present-day robot designs.

The idea that ASIMO needs somehow to seem friendly results in the way that this robot is designed to be slightly smaller than the average adult human and thus nonthreatening and also underlies the expressiveness of its communication. While ASIMO's designers have in the main concentrated their efforts on solving the problems of bipedal movement, they have also made ASIMO an adept mimic of many human hand and arm movements. ASIMO waves at people and shakes their hands, demonstrating how its humanlike form has been tailored towards supporting friendly communication with humans, as well as enabling its movement around the world. Although ASIMO's hand and arm movements show a tremendous potential for emotional expressiveness they are not overtly used to express emotional content. Indeed, there is no discussion of emotion as important in ASIMO's character on the Honda website, and in all versions its 'face' has remained immovable and partially obscured by a tinted visor. Therefore, although ASIMO may say things that contain some emotional content, its overall design demonstrates the idea that the expression of emotions is not that important to the operation of the robot or its interactions with humans.

Emotional robots

This raises the question of whether robots can, or should, express humanlike emotions in their communication with people. Even in Capek's play, the relationship between the conception of the robot, and the potential for its ability to feel, or to express emotions, was complicated. As the narrative of *R. U. R.* progresses, the robots are given the ability to sense pain, framed as a practical design decision in order to help them avoid damaging their fragile organic bodies as they work. However, towards the end of the play a range of emotions begin to emerge in the robot characters and, although the details are left rather unclear, it does transpire that a few robots were built with a 'level of irritability' as a way of 'making them into people' (Capek, 1920/2006, pp. 61–62). Unfortunately, while the robots' human designers may have meant to create more friendly robots, in the conclusion to *R. U. R.* the robots rebel against their human masters and it is their ability to relate to one another emotionally that allows the robots to survive and replace humanity in the world.

Asimov's fictional descriptions sometimes imbue robots with underlying feelings, although they often have difficulty expressing them as emotions to others. In 'The Bicentennial Man', for example, Asimov describes the robot Andrew's feelings as hidden because of the 'smooth blankness' of his face for much of the narrative (1976/1990, p. 245). It is only at the end of this short story that Asimov integrates Andrew's inner feelings with his expressions when upgrades to the robot's face make its features mobile, allowing the robot to smile (1976/1990, p. 290). It would seem that though Asimov's speculative essay 'The Friends We Make' does not overtly value the ability of robots to express emotions in their communication with people, in spite of positioning them as potential friends, his stories constantly question this idea. 'The Bicentennial Man', in particular, positions emotional expression in terms of body movement, tone of voice and facial expression as an important part of human-robot relations and communication.

The narratives of *R. U. R.* and 'The Bicentennial Man' highlight a difficult relationship between the robot as a constructed being, its ability to express emotions and its capacity to feel. The way that robots move, communicate and complete tasks places them as 'thinking' and also potentially as somewhat 'alive' but, as Turkle has noted is the case for computers (2005, p. 63), early in both stories humans are identified as unique because they have 'an emotional life' that robots lack. By the end of Capek's play and Asimov's short story this distinction is dissolved, the emotional development of Rossum's robots being driven by a human desire to make the robots more like people, whereas many of the changes to Andrew are chosen by the robot himself. In both cases, the robots are described as more richly communicative as a consequence.

It may therefore be unsurprising that some roboticists set aside the problems of perfecting bipedal movement to focus instead on maximising the expressiveness of the robot's face. Broadly, there are two different paths that these designs follow. The first leads to the creation of humanlike robotic faces that are virtually indistinguishable from humans, bringing Capek's artificial humans to mind. The second involves the creation of robots that operate more like caricatures of humans, a development path discussed later in this chapter. In terms of realistic-looking humanlike heads, the Japanese roboticist Hiroshi Ishiguro is famous for creating his own robot double, whereas David Hanson has created robotic heads that are accurate representations of Phillip K. Dick and Albert Einstein, as well as generic humanoids like Jules, built for the Bristol Robotics Laboratory (BRL) in the United Kingdom. Hanson explains that his aim is 'to model the behavior and movements of people in robots that act and react virtually indistinguishably from their human counterparts' (Hanson Robotics Website). For both Ishiguro and Hanson, the design of robots that communicate using human language and humanlike facial expressions is key.

Whether they have been designed to move in humanlike ways, or with humanlike expressive faces, very few real-life humanoid robots are sufficiently flexible and robust in their operation to enable their introduction into everyday situations. A few robots have entered the workplace, versions of Saya (a seated robot with a humanlike face) having worked as a receptionist, school teacher and greeting guests in a Tokyo retail store (Hornyak, 2009). ASIMO has also recently been trialled as a museum guide, although not entirely successfully (Hornyak, 2013). However, most humanoid robots are still found in laboratories and other spaces where people's interactions with them are carefully framed and controlled. It is therefore only in fiction that this type of humanlike robot has reached a level of sophistication that enables a more complete interrogation of the perceived benefits of such designs. In particular, Lieutenant Commander Data, the android officer in the television series *Star Trek: The Next Generation* (*ST:TNG*) and related films, provides a paradigmatic example of one possible future of this pathway in humanoid robotics.

Communicating with Data

Data is portrayed as a highly sophisticated humanoid machine, able to gesture and move in more humanlike ways than ASIMO, with a face that is capable of mimicking human expressions even more precisely than that of Jules. As his name suggests, Data is best regarded as a technologically advanced, embodied computer system. This robot's 'brain' is shown as composed of silicon chips and other electronic components, and he runs self-diagnostic programs when he feels there may be an error in his circuitry.

In terms of his appearance, Data is very humanlike indeed, although the makers of *ST:TNG* have defined some physical anomalies, such as his rather unusual eye colour and the metallic sheen to his skin, the latter being particularly noticeable in the first series. In addition, Data has some quirks in his behaviour that set him apart; for example, he rarely uses verbal contractions, saying 'is not' instead of 'isn't'. The difference between Data and humans most explored in many of the storylines is his lack of emotion, or more correctly the underlying feelings associated with emotional responses. Again, the discussion of Data's feelings resonates with Turkle's appraisal of the 'emotional life' that computers are understood to lack (2005, p. 63). However, Data, as an embodied computer with a humanlike face and body, is able to express the appropriate human emotional responses to different situations, although his inability to experience the related feelings continues to set him apart from the other crew members of the Enterprise.

For many of *ST:TNG*'s storylines, it is important that Data is almost indistinguishable from humans, while also remaining somewhat other-than-human in appearance, behaviour and ability. This subtle difference results in occasions when human characters are shown forgetting that Data is different from them in various ways. For example, in one episode Commander Riker makes several unsuccessful attempts to open a jammed door, before eventually stepping aside to allow Data, with his superior strength, to open the same door with ease ('Hide and Q', *ST:TNG*, Season 1, Episode 9). In addition, Data's human appearance and expressive face means that he often has to remind other characters that he does not have any feelings to hurt. In spite of Data's differences from humans, he experiences very few difficulties communicating with, working amongst and being accepted by the human, and humanoid alien, members of the ship's crew. Eventually, with the help of the 'emotion chip', Data is able to feel the effects of his emotions, a development that is portrayed as fraught with difficulty in its early stages as he learns to control the feelings that threaten to overwhelm his ability to function.

Most of Data's responsibilities in Star Fleet are concerned with research, information retrieval, processing and analysis, as might be expected in his role as Science Officer. An important part of such work, as is true for human scientists, is the ability to communicate results and conclusions to others. In general, Data seems to have few problems in this respect, although he is often unusually precise in what he says, using a systematic approach in his interactions and showing a particular concern to provide detailed and accurate information. The portrayal of his communicative acts stresses the importance of factual information, logic and objective rational argument. Data's communication style can therefore be analysed from the perspective of cybernetic and semiotic traditions of theory, which focus upon the accurate transmission or exchange of information, supported by precision coding and decoding in language.

The cybernetic tradition has its foundations in the work of scholars such as Claude Shannon and Warren Weaver (1948), Norbert Wiener (1948) and Alan Turing (1950), being drawn out of and feeding back into research into systems and information science, artificial intelligence and cognitive theory. The driving philosophical assumptions behind this tradition are concerned with materialism, rationalism and functionalism, the result being theories that regard all communication in terms of information processing and exchange within systems (Craig, 1999, p. 141). Of course, Data's construction as an embodied computer system serves to link his design with cybernetic theory, and with what is known as the cybernetic tradition of communication, even more closely.

As Craig notes, the cybernetic tradition shares some common ground with the other communication traditions he identifies, including the semiotic tradition. For Craig, this similarity is based on the way that semiotics collapses 'human agency into underlying or overarching symbol-processing systems' (1999, p. 141, citing Cherry, 1966; Eco, 1976; Wilden 1972). A similar link, between semiotic communication as a means to 'purge semantic dissonance' and cybernetic communication as a process of 'information exchange', is also made by Peters (1999, pp. 12 and 24). However, in order to explain how human communication as a process of information transfer or exchange actually takes place, it is useful to interweave semiotic and cybernetic theories of human communication even more closely than suggested by Craig or Peters. Semiotic theory describes human communication in terms of the use of particular signs whose meaning is shared between communicators. It can therefore be argued that it is semiotics, in particular the combination of signs that have precise intersubjective meanings in language, which enables information to be coded and decoded as part of a cybernetic process of information transmission or exchange. In discussing human communication, in particular communication with Data, a sophisticated humanlike machine, it may therefore be useful to identify a combined *cybernetic-semiotic* theory at work.

When communication is understood as this type of cybernetic-semiotic process, judging the success of communication becomes closely linked with measuring the accuracy in transferring meaning from one interlocutor to the other. Success in these terms therefore depends on the clarity of the information itself, the quality of its coding using intersubjective signs and the reduction of extraneous information, referred to as 'noise' in cybernetic theory. For some roboticists the development of humanoid robots, though complex and difficult, is therefore justified as a means of reducing the level of difference between humans and robots. The assumption is that this reduction of difference is of paramount importance in supporting more successful human-robot communication by enabling the use of familiar human codes or signs, including facial expressions and body movements. In spite of this, from the perspective of the semiotic tradition, even the ultimate humanoid robot, Data, shares the same problems with developing and using intersubjective signs as any human. *ST:TNG* storylines nevertheless position Data as a particularly formidable communicator, because he can learn any new language for which the grammatical rules and vocabulary are available.

Data's ability to collate, analyse and disseminate many different kinds of information accurately in familiar humanlike ways can therefore be understood to make him a near-perfect human-computer interface from the perspective of this cybernetic-semiotic model of communication. Data is positioned differently from Capek's robots, rather than being placed as a worker or a slave; his lack of emotion identifies him as the ultimate rational thinker and communicator within Star Fleet, an organisation for which these attributes are key. This is made particularly clear when Captain Picard, who strives to be a rational and unemotional commander, remarks that he wishes that he and the rest of the crew 'were all so well-balanced' as Data ('Datalore', *ST:TNG*, Season 1, Episode 12). For Picard, as well as the cybernetic-semiotic perspective identified above, emotions are clearly a form of 'noise', to be reduced as far as possible in order for successful information exchange to occur.

The concentration on the importance of rational thinking and reason over anything else in Data's communication also opens his communication to analysis from the critical tradition of communication theory. Theories within this tradition are concerned with identifying '[a]uthentic communication', suggesting that communication should have 'a built-in telos towards articulating, questioning, and transcending presuppositions that are judged to be untrue, dishonest or unjust' (Craig, 1999, p. 147). This sober view of communication as reasonable, productive discourse that results in coordinated action is linked with the work of Jürgen Habermas and his argument that only rational discourse can result in meaningful consensus (Habermas, 1987). Although decisions on the *Enterprise* may ultimately be the responsibility of Captain Picard, in regularly seeking the advice of his officers the importance of rational discourse, and in particular Data's role within this discourse, is also evident within the narratives.

Data's precision, and tendency to rely on reason as opposed to emotion, can be understood to produce a form of human language that has to an extent been 'machine coded' and therefore is fully separated from any underlying human feelings. However, Data's abilities still lead Craig to draw the conclusion that, from the perspective of the cybernetic, and I would argue also a cybernetic-semiotic tradition as well as the critical tradition, Data 'might be truly the most "human" member of the *Enterprise* crew', even in the absence of humanlike feelings (1999, p. 141). Thinking of Data in this way therefore validates the importance of reason, mind and language over emotion, body and nonverbal expression in humans for some perspectives on human communication. It also draws attention to the way that what is quite possibly Data's most important difference from humans – his lack of feelings – becomes the very aspect of his personality and communication style that makes him not only 'most "human"' but also the ultimate communicator (Craig, 1999, p. 141), at least within the context of his position in Star Fleet.

Avoiding the 'uncanny valley'

While Data is accepted by the crew of the Enterprise, and by many of the other humans and aliens that they meet, the same is not yet true of robots that have been designed to look as much like humans as possible in real life. For example, Jules remains a somewhat problematic humanoid, unnerving to catch sight of on a desk in the laboratory as I can personally attest having met Jules at the BRL. Having shown the video footage of Jules to a number of people, it is clear to me that this robot provokes a range of reactions. Some people are very impressed by the robot's facial expressions, but the majority are uncomfortable with the thought of spending time with Jules 'in person'.

The discomfort felt in the presence of robots such as Jules may be explained by the concept of the 'uncanny valley', a theory developed by roboticist Masahiro Mori that owes much to Sigmund Freud's definition of the uncanny as 'that class of the terrifying which leads back to something long known to us, once very familiar' (Freud, 1919/2004, p. 76). The theory is usually illustrated by a graph, which shows that as human likeness increases so does familiarity, up to a sharp drop-off point where a still doll is suddenly perceived as corpselike and a moving robot as zombielike (Mori, 1970). Mori's theory is considered by some to be particularly important in robot design, the effect being more pronounced for robots than for dolls. The prediction of this theory therefore is that a very humanlike robot such as Jules, because it still has small flaws in its appearance or behaviour that give it away, will be perceived by humans as eerily unpleasant, rather than attractive and interesting.

Only some researchers accept Mori's theory; others argue that there is as much evidence against the hypothesis as can be cited in its support. Indeed, Hanson, in justifying his design decision for Jules and other similarly humanlike robots, has proposed an alternative theory he calls the Path of Engagement (POE). Hanson argues that 'any level of realism can be socially engaging if one designs the aesthetic well' such that 'the illusion of life can be created and maintained' thereby mitigating 'the uncanny effects' (2006, p. 19). For Hanson then the problem is simply that Jules is not yet designed in such a way that the robot's aesthetic value consistently supports a valid POE, something which I am sure he would argue could be overcome in future designs.

However, while Hanson has no qualms about creating lifelike robots such as Jules, other roboticists have chosen instead to create robots in ways designed to avoid Mori's 'uncanny valley' completely. Cynthia Breazeal's robot, Kismet, for example, is clearly a machine, but nonetheless is able to mimic humanlike facial expressions (Figure 1.1). As such, Kismet demonstrates the possibilities offered by an alternative to realistic humanoid robot design, within which a cartoon-like face with stylised features is used to produce exaggerated humanlike expressions. Kismet presents as a table-top robotic head and neck with no body attached, as does Jules. However, the responses of human participants in videos of experiments with Kismet would seem to indicate that this robot does not produce an uncanny effect (Kismet Videos).

Communicating with Kismet

Kismet is sometimes referred to as a 'robotic creature', but is nonetheless most often positioned as a humanoid robot (Breazeal, 2002a, pp. 51 and 60). This is emphasised when Breazeal states that as a 'sociable robot' Kismet should be 'socially intelligent in a humanlike way' such that 'interacting with it is like interacting with another person' (2002a, p. 1). The highest goal for such robots is therefore that 'they could befriend us, as we could them' (Breazeal, 2002a, p. 1). She argues that since 'humanoid robots share a similar morphology with humans' they might well be 'capable of receiving, interpreting and reciprocating familiar social cues in the natural communication modalities of humans', and her research

FIGURE 1.1 *Kismet*
Source: Courtesy of Sam Ogden, photographer.

has been designed to investigate this idea (Breazeal, 2002b, p. 883). The similarities between Breazeal's goals for her robots and Hanson's, quoted earlier in this chapter, are therefore striking. In particular, both have a focus on building robots that are socially intelligent in such a way as to support friendships with humans.

Both of these streams of research, one producing near-realistic humanoid robots and the other producing robots with caricatures of human faces, are no longer simply concerned with enabling robots to operate in a physical world tailored for humans. Instead, the designs are based on the assumption that sharing humanlike expressions of emotion is essential to support human-robot sociability, which itself is linked with the perceived need to make robots that are able to communicate with humans in ways that are more subtle and emotionally aware.

However, while they aim to produce similar results in terms of sociability, the efforts of the design teams for Jules and Kismet have been differently focused. As mentioned earlier, in designing Jules, appearance has been of prime importance, and facial expressiveness helps to make this robot look as humanlike as possible. In contrast, Kismet's designers have simplified the task of producing expressions somewhat through the use of the cartoon-like face and have therefore been able to concentrate their efforts on programming Kismet to perceive the emotions of humans in conversation as well as responding in kind. Breazeal clarifies that her research focus in building Kismet was 'to explore dynamic, expressive, pre-linguistic, and relatively unconstrained face-to-face social interaction between a human and an anthropomorphic robot' (2002b, p. 883). She is therefore particularly interested in the passing of affective content via the prosodic cues contained in speech (tone, rhythm and intensity) and also using humanlike facial expressions.

In spite of the move away from creating a robot capable of producing well-developed human language, Breazeal's work still promotes the sense that the robot takes part in turn-taking dialogues. Kismet 'babbles' its turn, in the same way as a baby or a small child might (Breazeal, 2002b, pp. 884 and 892). This robot's design is therefore no longer based solely on the idea that a robot must be able to communicate effectively in cybernetic-semiotic terms, as was the case with Data. Instead, Kismet, as 'a sociable robot' (Breazeal, 2002, p. 1), may be better analysed in terms of theory that is concerned with social effects, whether individual or cultural, and therefore the sociopsychological and sociocultural traditions of communication.

Craig describes the sociopsychological tradition as theorising communication in terms of '*a process of expression, interaction, and influence*' (Craig, 1999, p. 143, italics in the original). For example, within this tradition, Lasswell (1948) models communication as a process about which researchers should be asking: 'Who? Says what? In which channel? To whom? With what effect?'. This understanding has much in common with cybernetic theories for which communication is a process of information transfer between a source and a receiver. The link is particularly clear when considering the 'hypodermic needle' theory of David Berlo, and also Wilbur Schramm's 'magic bullet' theory, both of which described the media as a source of communication, which was able to influence and alter the behaviour of individuals in an audience with ease. However, as Lasswell's description suggests, more recent thinking separates sociopsychological and cybernetic-semiotic ideas by involving a more careful analysis of how exactly successful communication is supported. In the main, the differences between the theories are concerned with the question of effect. Sociopsychological theory suggests that the reception of any message, and therefore its effect on the receiver, is mediated by psychological predispositions such as attitudes and beliefs, as well as the receiver's current emotional state. Thus the semiotic precision of the message and the reduction of extraneous noise are no longer enough to ensure successful communication. Instead, from a sociopsychological perspective the success of communication is also dependent on the message's ability to influence the receiver given their past experience and current state.

Data, whose communication focuses on cybernetic-semiotic excellence, demonstrates some ability to read social cues from others, but his reactions are shown as rather limited. In addition, while Jules might one day be regarded as a sociopsychological communicator, his current ability to read the communication of others for emotional content would seem to be restricted. It is therefore Kismet's design that most clearly draws on sociopsychological theory, acknowledging the importance of individual personalities, and the value of making an attempt to read the emotional cues of others as a vital aspect of communication before providing an appropriate emotional expression of one's own in response.

Unlike Jules, for Kismet, as a caricature of a human face, the problem of producing natural humanlike expressions is removed. The nature of Kismet's face means that humans interacting with it do not expect its expressions to be subtly nuanced or exactly like their own. However, Kismet is nonetheless perceived as needing to produce clearly recognisable humanlike expressions based on what is the most commonly identified set of 'basic' emotions: happiness, sadness, anger, fear, disgust and surprise, with the addition of a seventh, resting or neutral expression, for Kismet. This set of expressions is often identified with the work of Paul Ekman, amongst others, whose research supports the existence of a set of basic emotions that can be 'recognized from facial expressions by all human beings, regardless of their cultural background' (Russell, 1994, p. 102). The production of a set of clearly defined expressions is understood to make recognition of the expression as easy as possible for any human participant interacting with the robot. Kismet's design therefore illustrates a reliance on a different level of commonality from that of Jules. While not appearing to be realistically humanlike, Kismet's face is instead designed to mimic a basic and universally recognised set of expressions.

Experiments in which Kismet and human participants interact include those where humans are asked either to praise or to scold the robot, exaggerating their tones of voice as if Kismet was the infant it is designed to mimic. From video footage, available on the MIT website, Kismet would seem to be successful both in reading each human's intent and in influencing their subsequent response through its own exaggerated expressions of happiness on praise and sadness on scolding (Kismet Videos).

The above discussion highlights the importance of this robot's ability not only to express itself, but also to be able to perceive the affective states of those humans with which it interacts. It also draws attention to the way that this robot's success is in part supported by being given a particular position within a familiar human social structure: as an infant or a young child. The sociocultural tradition broadly describes communication as a 'symbolic process whereby reality is produced, maintained, repaired, and transformed' (Carey, 1992, p. 23). Communication within this tradition is regarded both as depending upon and reproducing the existing order of a social and cultural environment, and also as offering the possibility to produce new sociocultural patterns. Interactions with Kismet have clearly been designed to follow a particular relationship structure, in which Kismet takes the role of an infant, while the human participant is caregiver. Breazeal argues that this helps to make Kismet's limitations (in speech, nuanced expression and movement) more easily acceptable to humans (2002a, p. 51). It also encourages human participants to exaggerate their tones of voice and facial expressions, as they would when talking to a young child (Breazeal and Aryananda, 2002, p. 87). This exaggeration helps Kismet to pick up on the affective cues in the dialogue, allowing it to provide appropriate expressions in response.

The social and cultural positioning of Kismet has therefore been carefully chosen to support this robot's communication, and as the dialogues with Kismet unfold they continually maintain the infant-caregiver relationship. The fact that Kismet is clearly recognisable as a robot transforms the conception of this relationship somewhat, by demonstrating that the participants in such exchanges need not be human, but it does still suggest that the robot needs to be humanlike. Indeed, this analysis of Kismet's design and communication is both based on, and reinforces, the way in which these social conceptions of communication advocate the need for commonality between human and robot in order to support successful interactions.

Critiquing the pursuit of sameness

A number of science fiction authors and real-life roboticists embrace the assumption that humanlike form, including in many cases the provision of an expressive humanlike face, is the most effective way to support working and social relations between humans and robots. This assumption can be related to understandings of communication that place an emphasis on the idea that some level of commonality between interlocutors is essential for successful communication to occur. Placing value on commonality, for example, is often a part of seeing communication in terms of cybernetic-semiotic processes of information transfer or exchange, critical understandings of the importance of rational discourse, sociopsychological attempts to persuade or influence others, or sociocultural conceptions of the production and reproduction of shared social understandings about the world. In addition, the theories of communication I've discussed in this chapter – cybernetic, semiotic, critical, sociocultural and sociopsychological – often support one another in reinforcing the idea of the robot as like another human in a particular context. Then, by valorising the importance of commonality in supporting a vision of communication success, the creation of humanoid robots also perpetuates specific understandings of the communication traditions discussed here.

Though not much talked about in robotics, issues with the tendency to value what communicators have in common over and above their differences, and the different perspectives they might have to offer, have received attention within communication scholarship. Peters notes that the philosophy of Emmanuel Levinas, one of 'the most influential thinkers about communication' in recent times, strongly supports the value of difference in communication (1999, p. 20). In particular, Levinas argues that when communication is 'sought as a fusion', such that '[o]ne begins with the idea that duality must be transformed into unity, and that social relations must culminate in communion', it comes to bear 'the mark of failure or inauthenticity' (1989b, p. 164). Peters draws on this appraisal to critique 'the dream of communication as the mutual communion of souls' and the 'pervasive sense that communication is always breaking down' (1999, p. 1). In particular, he argues that if communication is 'taken as the reduplication of the self (or its thoughts) in the other' then it 'deserves to crash, for such an understanding is in essence a pogrom', an organised massacre, 'against the distinctness of human beings' (Peters, 1999, p. 21).

Peters is not the only communications scholar to question the way in which basing understandings of communication theory on commonality leaves little space for the other. Pinchevski (2005, p. 65) offers a precise critique of four 'traditional conceptions of communication' that also draws on the philosophy of Levinas. Although his categories are different from Craig's seven traditions, the two frameworks are reasonably easy to link, in particular if the combined cybernetic-semiotic model suggested earlier in this chapter is adopted. Pinchevski's arguments are specifically focused on clarifying the 'conceptual blindness imposed by an unwillingness or inability to acknowledge otherness within the gamut of communicational phenomena' (2005, p. 29). He therefore explores how traditional theories' descriptions of communication can be understood to support the elimination of otherness and difference, resulting in a level of 'violence' to the other (Pinchevski, 2005, p. 29–65).

In summary, Pinchevski sees violence arising from traditional communication theory whether this theory is associated with the clarity of a system's communication channels and the successful transmission of the message sender's intent (cybernetic-semiotic); the insistence that rational discourse is the only path to agreement (critical); the use of persuasion to manipulate and force another to take a particular point of view (sociopsychological); or, the idea that communication offers a means to include or exclude someone from a social group or a society (sociocultural) (2005, pp. 29–62). His overall appraisal of these traditions of communication theory is that they 'are largely about the reduction of difference or the transcendence of difference' and the eventual 'elimination of difference' (Pinchevski, 2005, p. 65). The idea of communication as 'a great equalizer' is viewed from within any of these theoretical perspectives as highly beneficial (Pinchevski, 2005, p. 65). However, Pinchevski does not agree, and through his specific critiques of traditional communication theories he comes to a similar conclusion to that drawn out by Peters' more historical argument, that all of these 'traditional' understandings of communication result in an unethical level of violence against the other, involving disrespect for them and their personal, cultural and social differences from the self.

As this chapter has shown, analysing various different forms of humanoid robot as reduplications of the self, or in terms of eliminating difference, at various levels is certainly possible. Its effect may not be so important as the violence against the human other, but the creation of familiar robots, designed to support easy nonchallenging communication, may not only reduce the possibilities for humans to interact with a wide variety of forms of robot, but also limit the new perceptual skills and abilities that robots might bring to bear in human-robot teams. This is clarified within the explanations of roboticists such as Breazeal, and her doctoral supervisor at MIT, Rodney Brooks, relating to the design of robots that they describe as '*situated*' (2003, p. 51). As opposed to relying upon fixed internal models of the environment, their robots are 'embedded in the world' in such a way that a real-time ability to perceive their surroundings directly influences their behaviour (Brooks, 2003, p. 51). Brooks argues that making a robot humanoid allows 'humans to interact socially with the robot in a natural way' (Brooks et al., 1999, p. 57), an idea both complemented and extended by the authors of a paper reviewing a large number of varied robot designs who note that '[t]o interact meaningfully with humans, social robots must be able to perceive the world as humans do' (Fong, Nourbakhsh and Dautenhahn, 2003, p. 155). Therefore, while robots could be situated in the world in novel ways and thus be able to offer nonhuman perceptions of the world in working partnerships, a number of roboticists focus instead on creating robots that are physically, through their form and their senses, and socially, through their faces and expressions, situated in as similar a way to humans as possible.

The argument developed throughout much of the rest of this book demonstrates that the provision of humanlike form, senses and emotional expressions is not the only way to build robots that can 'interact meaningfully' or 'interact socially...in a natural way' with humans (Fong, Nourbakhsh and Dautenhahn, 2003, p. 155; Brooks et al., 1999, p. 57). My focus is on suggesting that rather than concentrating on the need for commonality, and the idea that communication is solely founded in the similarities between what are often diverse individuals, difference is an essential ingredient in all self-other relations. By allowing space for difference in communication it is possible to recognise that numerous situated knowledges about the world, including nonhuman ones, exist and are valuable (Haraway, 1988). Chapter 2 begins this process with an exploration of human-animal communication, as a basis for a return to considering human-robot interactions in a new light in the chapters that follow in Part II of the book.

2
Human-Animal Communication

Abstract: *Chapter 2 pauses the book's consideration of robots to explore human-animal communication, since nonhuman animals are an important part of many people's lives, acting for some not only as companions, but also as co-workers. Human-animal communication is described in various, often idealised, ways in fiction, but in real-life situations an analysis of human-dog communication demonstrates the importance of attending to the smallest of nonverbal signs over periods of dynamic communication. This chapter highlights the possibilities for humans and animals, in particular dogs, to work together in teams, employing their specific skills to allow the team to perform tasks that neither human nor dog could complete alone. At the end of the chapter is a brief appraisal of the design and development of animal-like robots.*

Sandry, Eleanor. *Robots and Communication*. Basingstoke: Palgrave Macmillan, 2015. DOI: 10.1057/9781137468376.0007.

The argument that humanlike form is a requirement to create robots that can work in human-tailored environments, are easy for humans to communicate with and potentially familiar enough to be perceived as friends overlooks the fact that many humans interact on a regular basis with nonhuman others in the form of animals. Many of these animal others don't just exist alongside humans; rather, they interact with humans directly and, at times, communicate with them such that humans and animals work together to complete tasks that neither could accomplish on their own. This chapter therefore considers various examples of human-animal relations and interactions, including those that illustrate the development of effective working teams. An analysis of communication across species identifies new perspectives on communication and opens up the possibilities of developing non-humanoid robots to collaborate with people in a variety of situations.

Animal communication in fiction

Before moving on to consider real-life interactions between humans and animals, it is worth noting that in fiction, human-animal communications often occur in idealised ways that appear to allow even very complex interactions to proceed with relative ease, supporting narrative development and audience understanding. Animals are sometimes portrayed as possessing the ability to talk in human language with one another, although they don't always use this capability to communicate with humans. In the film *Babe* (1995), for example, all the farmyard animals talk to one another, but not with the farmer Arthur Hoggett. The film's characterisation of the animals and the progression of its narrative is reliant on the animals' ability to speak, with sophisticated interactions between species supported by the nuances afforded by the tones of a human voice and humanlike expressions transposed onto the faces of the animal characters. Although the animals speak to the audience, the idea that language-use forms a boundary between humans and animals remains intact within the film, because the animals do not speak to the farmer. In this type of narrative construction, also a staple of Disney animations, the communicative abilities of animals in real life are replaced by a fictitious ability to use human language, tones of voice and facial expressions. The animal's otherness is reduced as it is made as humanlike as possible in order to support the audience's understanding of the plot.

Other animal characters, famous in film and on television, such as Lassie (the dog), and Flipper (the dolphin), cannot speak using human language; however, these animals are shown to have very few problems communicating with people using other means. The animal actors that star in these films and programmes are, of course, highly trained and responsive to human commands in real life. It is therefore reasonable for fictional narratives to portray them as understanding human voice commands and complex signals with ease, although at times the novelty of the situation being faced would seem likely to stretch even the most well-trained and attentive animal. However, the narratives also show these animals in situations where they not only attract people's attention, but also explain seemingly complex occurrences, when, for example, a storyline requires them to lead human rescuers to the site of an accident. The communication between human and animal follows the form of a dialogue, one side producing signals using movement and dog barks or dolphin whistles, while the other talks. In these situations, the actions of the animals are normally 'translated' for the audience via the commentary provided by human characters. This narrative technique has resulted in a set of paradigmatic misquotes used to summarise interactions with Lassie, as seen in an episode of the popular television series *The Big Bang Theory* (*TBBT*): 'What's wrong Lassie? Timmy fall down the well?' where Penny is making fun of Raj, tongue-tied as usual in the presence of a pretty girl ('The White Asparagus Triangulation', *TBBT*, Season 2, Episode 9). Although this exact line was never uttered in the films or television programmes starring Lassie, it does sum up the way in which Lassie's communicative ability and care for humans is generally remembered.

These various portrayals of animal communication in films and on television are shaped by the need to provide viewers with a clear plot narrative, so that the storyline is evident to the audience. However, they also serve to occlude people's recognition of animals as others with their own specific communicative abilities. Even Lassie and Flipper are positioned as needing to understand human language, both verbal and nonverbal, as well as interpreting situations *as another human would*. The animals are then required to respond in ways that seem immediately understandable to human characters, in order to form partnerships with people and to complete shared tasks effectively. As I am about to discuss in more detail, real-life communication between humans and animals often proceeds not as a turn-taking dialogue, but rather as a form of dynamic system within which small signals of movement and sound overlap and build upon one another. At times, the meaning of a communicative act may immediately be understood, as in the case of a warning bark, for example. On other occasions, the precise meaning of each communicative act is less clear, and the dynamic system instead supports the gradual emergence of meaning between human and animal over time as these events overlap and combine.

Communicating with Clever Hans

A reappraisal of animal communication with humans requires a rethink of what communication entails, because human perspectives on what constitutes communication are often not that helpful when trying to understand the communication skills of animals. This issue can be illustrated by analysing the real-life story of Clever Hans, a horse that was trained to communicate by tapping his hoof on the ground. Hans was first taught to count out a number, and then to tap out the answers to mathematical calculations. As his training progressed, Hans was even able to use an alphabet coded into hoof taps to answer verbal or written questions (Heyn, 1904). Clever Hans was therefore thought to be a remarkable horse, proclaimed as showing 'an intelligence beyond that of many human beings' because of his numeric and linguistic abilities (Knapp and Hall, 2010, p. 3).

His status was reconsidered when it was discovered that Hans was not really able to count, perform arithmetic or answer other questions. In fact, Hans was paying close attention to the body language of people around him and reading their nonverbal signals as a means to know when they wanted him to stop tapping his hoof. His trainer and the crowd were completely unaware that they were providing Hans with the correct answers through their spontaneously produced nonverbal cues. The misunderstanding over Hans' abilities was finally uncovered when an experimenter and the trainer each whispered numbers to the horse. Since neither the trainer, nor the experimenter, nor anyone in the audience knew both numbers, and therefore the answer to the sum, there were no surrounding nonverbal cues to tell Hans when to stop tapping and he was unable to give the correct response (Knapp and Hall, 2010, pp. 3–4).

There are a number of ways in which this story can be interpreted. It is, for example, tempting simply to note that Hans was not so clever after all, unable to count or to answer questions on his own without human cues to give him the answer. From this perspective, 'the Clever Hans syndrome' becomes a 'cautionary tale in animal communication research about the dangers of projecting ourselves onto animal subjects' (Peters, 1999, p. 242). The story can also be understood as 'the classic scenario of miscommunication: a smooth interaction that, it turns out, has radically different meanings for each participant' (Peters, 1999, p. 242). From this perspective, it is a reminder that what is actually communicated can often be radically different from the intended message (Knapp and Hall, 2010). However, it is also possible to view the situation as an illustration of excellence in nonverbal communication, admittedly not planned but rather emergent. In particular, the story draws attention to the 'glances, gestures' and 'positionings' as well as the 'verbal statements' that are constantly produced by humans, 'whether intended or not', which Erving Goffman describes as 'small behaviors' (1972, p. 1). The Clever Hans story therefore highlights how nonverbal cues are valuable aspects of communication and, while they are sometimes easily overlooked, one can learn to read them, whether one is a horse, a human or potentially something entirely other.

A reconsideration of this story from a perspective that acknowledges Hans as 'clever' because of his excellence in reading human body language goes some way to indicate the possibilities of human communication with overtly different others. In particular, it introduces the importance of paying very close attention to the body of the other in communicative situations. Of course, this example doesn't allow me to explore in any real sense how communication between humans and animals occurs. While teamwork was required to complete the task, the humans involved were unaware that it was their abilities to count, do arithmetic and answer questions, and the resulting body language signals they unknowingly produced, that allowed Hans to provide the correct answers. It might even be possible to argue that in this situation, Hans was the better communicator, noticing the nonverbal communication of humans, acting on this in a way that was clear for his trainer and other human onlookers, and recognising the reward of doing so!

What is clear from this story is that Hans is relatively easy to categorise as a special horse, as a defined individual with particular skills, even after his abilities have been reassessed not as mental arithmetic but as excellence in reading human body language as a cue for action. This acknowledgement of an individual animal, and the decision to pay close attention to them, allows people to become more aware of the specific ways in which animals communicate, as well as the human verbal and nonverbal signals that they themselves are contributing to the interaction.

Recognising individual animals

The importance of attending to an animal's behaviours, and at the same time acknowledging its ability to attend to the behaviours of humans, is highlighted in the work of Barbara Smuts. She suggests that in order to communicate with animals it is necessary to 'recognize and relate to the animal as an individual'; as she goes on to explain, 'to use "the philosopher's idiom", the animal must get a face' (Smuts, 2008a, p. 125 note 17). Here, Smuts is referring not to a physical face, but rather to a philosophical conception within which the appearance of a 'face' denotes the expressive and potentially responsive presence of another being.

While acknowledging an animal's 'face' is essential to Smuts' understanding of her relationship with her dogs, attaining this level of recognition for the animal has not proved so easy for the majority of Western philosophers. In 'The Animal that Therefore I Am (More to Follow)', Jacques Derrida explores his dissatisfaction with the philosophy of 'Descartes, Kant, Heidegger, Lacan and Levinas' whose gaze seems never to have 'intersected with that of an animal', such that they take 'no account of the fact that what they call animal could *look* at them and *address* them [...] from a wholly other origin' (2002, p. 382). For example, in an interview, Levinas says that, while '[o]ne cannot entirely refuse the face of an animal', in animals '[t]he phenomenon of the face is not in its purest form' (Wright et al., 1988, p. 169). However, in spite of making this statement, there are moments in his writings which indicate the complex nature of the question of the animal's 'face'.

The issue is particularly clear in his description of Bobby, the stray dog that enriched his life and the lives of other prisoners of war incarcerated with him in Nazi Germany. While all the people with whom the prisoners came into contact treated them as subhuman, Bobby 'would appear at morning assembly and was waiting for [them] as [they] returned, jumping up and down and barking in delight' (Levinas, 1990, p. 153). Levinas suggests that for Bobby, unlike the humans that dealt with the prisoners, 'there was no doubt that we were men' (1990, p. 153). Nevertheless, Levinas still argues that Bobby has 'neither ethics nor *logos*' (1990, p. 152).

As David Clark notes, 'Levinas' experience of Bobby is informed by conventional assumptions about animality that make it impossible for him straightforwardly to attribute dutifulness to a creature that is not human' (1997, pp. 165–166). It should also be acknowledged that Levinas' account is focused on establishing an ethical point relating to the treatment of human prisoners, and his argument involves differentiating them from all other beings.

Therefore, in spite of the way that the dog Bobby provided a respite from the inhumanity of life in the camp, it was difficult for Levinas to suggest, given the context about which he was writing, that Bobby should be accepted as a being. However, Clark offers a compelling alternative to Levinas' conception of Bobby when he asks 'what is "language" if it is not the wagging of a tail, and "ethics" if it is not the ability to greet one other and to dwell together *as* others?' (1997, pp. 190–191). For Clark, Bobby clearly does have a 'face' that is revealed through his nonverbal behaviour as he greets the prisoners on their return at the end of the day.

Derrida's own experience one morning, 'naked under the gaze of a cat', is 'that nothing will ever have done more to make [him] think through this absolute alterity of the neighbour' than this encounter (2002, p. 380). In a similar response to that of Smuts, Derrida accepts this specific '*little cat*' as having a 'face' (2002, p. 374, italics in the original); although, as Donna Haraway points out, he appears to learn 'nothing more *from, about, and with* the cat' having recognised that he was under its gaze (2008b, p. 22, italics in the original). Instead of asking 'what this cat on this morning cared about, what these bodily postures and visual entanglements might mean and might invite', his analysis concentrates 'on his shame in being naked before this cat' (Haraway, 2008b, p. 22). Derrida therefore doesn't seem to take full advantage of this opportunity to communicate with the cat, although his recognition of its 'face' does open the way for such an interaction to occur.

Smuts, however, not only recognises the gaze of animals, but also argues that individual animals have 'idiosyncratic, subjective' experiences of humans that combine with human appraisals of them to support meaningful human-animal interactions (2001, p. 118). In a personal account of her relationship with her dogs Safi and Bahati, Smuts describes how, 'instead of issuing "commands"', she 'experimented with simply talking to Safi conversationally, as [she] would with another human' (2008b, p. 119). She notes that Safi 'developed the habit of looking inquiringly at [her] when [they] encountered novel situations' and talks about the way that this dog learned to 'trust' her, leading to the development of a 'mutually respectful relationship' between dog and human (Smuts, 2008b, pp. 119–120). Smuts also describes her other dog, Bahati, as capable of acting 'politely' when 'trying to get [her] attention' (2008b, p. 120).

Reading these descriptions of communication with her dogs, it is easy to assume that Smuts is anthropomorphising the dogs' behaviours, attributing human abilities and responses onto them through her interpretations of their nonverbal communication. However, she would not agree. Instead, she explains that when she experiences 'individuals of many other species first and foremost as *persons*' her decision is not related to attributing 'human characteristics to them' (2001, p. 118). From Smuts' perspective, her responses to dogs and her reading of their responses to her are not anthropomorphic, because she recognises 'that they are social subjects' in their own right (2001, p. 118). In particular, Smuts is concerned to acknowledge the social skills of animals, arguing that being able to understand elements of human language, inquire, trust, respect and be polite are actions that dogs, and other animals such as primates discussed in her broader research, can perform just as well as humans.

Smuts describes relationships with her dogs as 'a perpetual improvisational dance, co-created and emergent, simultaneously reflecting who we are and bringing into being who we will become' (2008a, p. 115). This idea is also picked up by Haraway, who suggests that the communication Smuts describes, not only in discussing relations with her dogs, but also in her observations of greeting rituals in baboon society, 'is more like a dance than a word' (Haraway, 2008a, p. 111). In exploring this idea of communication as a form of dance, Smuts draws upon the work of Stuart Shanker and Barbara King, who suggest that the 'dance metaphor', and the 'dynamic systems paradigm' that underpins this description of communication, have become prevalent in a number of areas of communication research, citing ape language research, nonverbal communication research and infant development research in particular (2002, p. 605). Shanker and King identify the central concerns of the dance metaphor as '*co-regulated* interactions and the emergence of creative communicative behaviors within that context' (2002, p. 605, italics in the original). This perspective on communication therefore concentrates more on the interaction itself, and the development of the relationship that occurs during that interaction, than on the individual participants and their actions.

Communicating about relationships

What has not yet been clarified in this discussion is what it is that can be said to emerge from the dances between participants. In her own analysis Smuts describes the '*embodied communication*' that takes place with her dogs as being a critical part of an 'inter-species language' (2008b, pp. 136–137, italics in the original). While 'speech sometimes plays a role', her experience and the research of others 'suggests that dogs often respond more to human body language and tone of voice than to the words themselves' (Smuts, 2008b, p. 137). The meaning of such interactions 'does not reside in the specific behaviours shown, nor does the interaction refer to something "out there" in the world' (Smuts, 2008b, p. 137). Instead, Smuts argues that 'meaning is mutually constituted, literally embodied as two individuals' behaviours ("the parts") combine to create something new ("the whole")' (2008b, p. 137). Smuts' explanation positions communication that is embodied in this way as referring 'to interactions whose meaning lies more in such *emergent properties* than in the lower-level, individual actions of the participants' (2008b, pp. 137–138, italics in the original). Attending to the complexity of nonverbal communication as it flows between participants therefore draws attention to the meaning that emerges within a system of communication itself.

Smuts' ideas also focus on this communication being about the relationship, as opposed to be about something else 'in the world' (Smuts, 2008b, p. 137). In contrast, while parts of Haraway's writing seem to value the relation above all else, in other places she clarifies the importance of the history and situatedness of the participants, as well as the value of the relationships they enact. For example, she suggests that 'actors become who they are *in the dance of relating*, not from scratch, not ex nihilo, but full of the patterns of their sometimes-joined, sometimes-separate heritages both before and lateral to *this* encounter' (Haraway, 2008b, p. 25). The presence of the participants as individuals is made clear in this description, since they are not created in the relation, rather the relation is filled with them, their histories and their experience.

Companion species and communicating about what's 'out there'

The discussion of communication above is concerned mainly with how attending to embodied communication helps to explain the ways in which relations are developed through interaction between humans and animals. Smuts' analysis of her relations with dogs has drawn out the possibilities of regarding the animal other in these interactions in some sense as a person, in terms of being socially aware, without considering them to be human. This type of approach encourages theorists such as Haraway to argue that interactions between humans and animals can be rich, functional and valuable. In particular, this perspective is to be found in her conception of 'companion species' and the possibilities of their relationships with humans (Haraway, 2003). Haraway develops her conception of the term companion species in part through a consideration of her own relationship with her dog Cayenne as they live together, while also training and competing in dog agility trials.

Dog agility is like an assault course for dogs. Each competition consists of a set of obstacles, over and through which the dogs run, jump and weave, adhering to certain rules and with the aim of completing the course in the fastest time. Only the human members of the team see and walk the course before the contest. While the dogs learn what is required to complete each particular obstacle, through training and repetition at home and in low-level competitions, they are completely reliant on the human team member for directions, in order to take all obstacles at speed, in the correct order and direction for each new course they are asked to run.

During a competitive run, only a few words are used as direct commands from human to dog. Much of the communication depends on the 'glances, gestures and positionings' that Goffman terms 'small behaviors' in considering human interactions, so effectively read by Clever Hans and overlooked by humans (1972, p. 1). In the team situation of dog agility, however, regular glances between dog and human ensure that their body positions and movements relative to each other and to the obstacles on the course are continually monitored. Communication in a dog agility team can therefore be regarded as involving the continuous production and reading of 'small behaviors' by both human and dog (Goffman, 1972, p. 1). It is these small signals, expressing understanding and intent, or sometimes confusion needing reassurance about what must be done next, that drive the team's capacity to complete the course quickly and correctly. Communication operates as a dynamic system during this type of embodied communicative situation, and signals between communicators overlap as human and dog continually reassess each other's position, perceived intention and likely subsequent action.

When watching dog agility contests the 'dance metaphor' seems an immediately appropriate way to describe the communication that takes place within the team as human and dog move around the course (Shanker and King, 2002, p. 605). Importantly, this communication is not only about their relationship, but also concerns the course that is being run. It is therefore about something 'out there' in the world, contrary to Smuts' contention that this type of embodied communication is all about relation (2008b, pp. 136–137). In spite of this, the development of the relationship and taking time to learn each other's communicative styles is essential to the final success of the team in competition.

Haraway suggests that it is by taking part in training and competitions over time that she and Cayenne have become '*constitutively, companion species*', a relation within which they have become '*significantly other to each other, in specific difference*' (2003, pp. 2 and 3, italics in the original). The development of this relationship involves complex and dynamic interactions that allow dog and human to end up '*training each other in acts of communication [they] barely understand*' (2003, p. 2, italics in the original). Haraway's statement draws attention to the presence of 'small behaviors', alongside the sense in which the ability to read those signs takes a practice. Training together is not just about learning the technical requirements to complete the various elements on the course: it is also an essential part of learning how to communicate across the significant difference between species in order to work well together as a team.

Learning to value anthropomorphism

Smuts' and Haraway's descriptions of communication with dogs resonate with the writing of Vikki Hearne, another researcher into human-animal relations, who openly acknowledges how anthropomorphism, and the descriptions of animal behaviour that it supports, can help humans and animals to work together more effectively. Hearne's research was driven by her interest in the 'highly anthropomorphic, morally loaded language' that trainers use to talk about animals, because it seemed to be 'part of what enabled the good trainers to do so much more than the academic psychologists could in the way of eliciting interesting behavior from animals' (2000, p. 6). Hearne argues that communication between humans and animals is far more effective when people use anthropo-morphism in thinking about their animal's behaviour than when they refuse to allow anthropomorphic responses to play a part. In spite of this, as she and others have noted, describing an animal as honest or kind, or recognising an animal's ability to take responsibility for the completion of a task, is often frowned upon as undermining the objective stance that scientists strive to attain (Hearne, 2000; Bekoff, 2007; Flynn, 2008). From Hearne's perspective, stating that a working dog knows her job and is thus capable of being 'responsible' is simply the use of a terminology that, while it may be based in unapologetic anthropomorphism, is the only form of description 'that *makes sense* of the situation' when, for example, working with an experienced scent-tracking dog (2000, pp. 9 and 14, italics in the original).

At the same time Hearne recognises how difficult it can be to trust that a dog is taking responsibility for the job at hand, because their assessment of the situation is based on different senses from that of the human. Tracking scent trails is very different from looking for other physical marks, such as footprints or broken foliage, because the scent trail drifts with the wind over time. In training situations this becomes evident when the human knows where the trail has been laid and becomes impatient with the dog that is following its nose along an arc away from that trail (although nonetheless tracking very effectively if left to complete the task). In addition, other factors can come into play causing misunderstandings of the dog's actions. In particular, Hearne describes an occasion when she was training the dog, Gunner, on a scent trail that had been laid by her daughter Colleen. Hearne explains that dogs are not only trained 'to follow a scent', but also 'to retrieve objects dropped by the track layer' (2000, p. 13). On this occasion the track was quite clear to Hearne, because dew on the ground meant that the path Colleen had taken was visibly marked. Therefore, as 'Gunner abandoned the trail and began bounding to the left, toward some bushes' Hearne felt justified in angrily trying to stop him from making this detour, assuming that she 'knew more than the dog about what was going on' (2000, p. 14). It was only when he emerged with a toy that Colleen had lost a few days before that Hearne realised Gunner had proved her wrong, and in the process reminded her that dogs are capable of taking responsibility for the completion of a task according to their training (2000, p. 14). Hearne's story highlights the way in which training and learning together is an important aspect of any interspecies team, because it is only through practice that team members can learn to trust each other's abilities and commitment to the task at hand.

From human-animal communication to understanding robots

In relation to this book's concern with human-robot interactions, an understanding of animals as social and communicative in complex ways is important because it supports the possibilities of zoomorphism, as well as anthropomorphism, in analysing human interactions with machines. As discussed in Chapter 1, researchers designing robots to interact with humans, such as Breazeal and Hanson, tend to assess social ability as predominantly a human trait, and thus they overtly frame their robots as anthropomorphic, in order to increase the level of anthropomorphism that their robots support. However, if one embraces the possibilities of animals as communicative and collaborative others, as Hearne, Smuts and Haraway clearly do, then zoomorphism becomes a valuable means of interpreting the attributes and behaviours of non-humanoid robots in social terms. From this perspective, it would seem possible to argue that, while the personality and character of robots is an important part of the way in which people interact with them, this personality and character need not be associated only with humanlike form.

Some roboticists have followed through with the idea that pursuing animal-like design will support human interactions with their robots. This was seen with AIBO, Sony's robotic dog (now discontinued), and Paro the robotic seal, a robot designed as a health care and therapy tool. While these robots are not exactly like the animals they resemble, their behaviours are based on those of animals. AIBO, for example, was designed to learn and respond to voice commands like a puppy, whereas Paro responds to stroking not so much in a seal-like way (because not many people have stroked a seal) but certainly in recognisably pet-like ways. AIBO and Paro were both designed to mimic human-animal relations, with Paro's creation aimed at providing a comforting companion for people living in spaces where live animals cannot be introduced. These animal-like robots are attractive and nonthreatening, supporting simple and easy-to-understand interactions with which people might already be somewhat familiar based on their prior experience with pets or other animals (Leite, Martinho and Paiva, 2013). This design decision, while not pursuing the idea of commonality in the same way as the creation of humanlike robots, follows a similar course in creating robotic animals. As such, it suffers from some of the same difficulties, notably the limitations imposed on how the robot can behave, curtailing the unusual attributes that it might bring to the human-robot relation.

Communication with AIBO and Paro is designed with a stimulus response model in mind and therefore employs the type of behavioural training assumptions and techniques about which Hearne raises questions. The interactions with these robots, designed to proceed in ways that are easy for humans to understand, become stereotypical versions of human-pet relations. These robots do not draw attention to the otherness of animals and the possibilities of employing the nuances of dynamic communication and 'small behaviors'. These designs cannot therefore result in the creation of human-robot teams that reach the potential more developed human-animal relations show, in particular where humans and animals train together in ways that enable their individual skills, abilities and communicative styles to contribute to the performance of the team in completing collaborative tasks.

The next part contains a set of chapters that explore a number of alternative robot designs, from those that support only relatively simple interactions with humans to those where the interaction is more sophisticated and allows human and robot to work together in a collaborative team. All of these robots are clearly non-humanoid, and broadly also not animal-like. In spite of this, they are often interpreted as somewhat like people or like animals, but these anthropomorphic and zoomorphic responses are always positioned against the clarity with which the robots are also understood to be unusual others that do not overtly look at all like humans or animals. The marked strangeness of these robots leads to the suggestion that interactions with their radical otherness offer the chance to come into contact with many new perspectives from which to perceive and understand the world.

Part II
Communicating with Non-Humanoid Robots

3
Encountering Otherness

Abstract: *Chapter 3 concentrates on theorising the encounter between human and robot, identifying moments when communication occurs, often using nonverbal communication channels at least initially. It discusses two versions of the Autonomous Light Air Vessels (ALAVs) art installation, within which blimp-like robots interact with one another and with visitors. It also extends Levinas' conception of self-other encounters to consider nonhumans, including robots, in order to consider how communication can be understood to draw the self and other into proximity while retaining the differences between them.*

Sandry, Eleanor. *Robots and Communication*. Basingstoke: Palgrave Macmillan, 2015. DOI: 10.1057/9781137468376.0009.

Although, as discussed in the first chapter, some roboticists argue that in order to interact with people robots need in some way to be humanlike in form and communicative ability, not all designers embrace this perspective. While others have explored the design of animal-like robots, as briefly discussed at the end of Chapter 2, it is often the creations of artists that really push the boundaries of what is possible when humans interact with strange robots that are not humanlike or animal-like in appearance. This chapter therefore considers two versions of the Autonomous Light Air Vessels (ALAVs) installation, created by artists Jed Berk and Nikhil Mitter. The different iterations of this project illustrate a number of forms of communication as humans and these strange robots interact. Thinking through interactions with and between ALAVs encourages development of a broader understanding of the possibilities of communication that draws upon and further extends the ideas already introduced in relation to human-animal interactions.

The Autonomous Light Air Vessels

The ALAVs are small groups of flying robots that interact with one another and with human visitors. Each robot is constructed from a blimp-shaped helium-filled balloon, with an electronic controller and a pair of small propellers mounted underneath (Figure 3.1). While the basic nature of their behaviours and actions are programmed, the ALAVs decide when to produce certain sounds and movements based on their own real-time perceptions of the surrounding world.

In version 1.0, the ALAVs consisted of a small flock of three robots named Flipper, Bubba and Habib. Human visitors and the ALAVs moved freely around one another within the installation space, supporting Berk and Mitter's aim to develop 'an interactive system' to which people are introduced 'as part of the ecology' (ALAVs Website). The ALAVs roam around searching their surroundings for food while remaining constantly aware of one another's positions at all times. If a robot becomes isolated for too long it calls out, making a sound produced by a mobile phone vibrator amplified against its helium-filled balloon body. This echoing sound is reminiscent of whale song and is described as a 'nervous shout' by Berk and Mitter (ALAVs Website). In addition to engaging in a search for 'food', these robots are described as 'yearning for attention', and on meeting after some time apart they perform a 'flocking behaviour' during which each robot spins around and makes repeated short calls (ALAVs Website). The behaviour of these robots in version 1.0 of the installation encouraged human visitors to interpret them as being a connected group of inquisitive individuals, willing to explore their environment separately, while remaining constantly aware of the group as a whole.

People are also able to interact directly with the ALAVs by holding a fibre optic lamp, which acts as the robots' 'food source' (as shown in the right hand side of Figure 3.1). The food source attracts the ALAVs, bringing them closer to their visitors and encouraging an understanding of the ALAVs as creatures. Feeding the ALAVs is rather like taking birdseed into an aviary or sugar solution into a butterfly enclosure; however, the details of the process make it somewhat unusual and unique. Once positioned above the lamp, an ALAV indicates that it is feeding by flashing the blue light on its underside in time with the lamp food source, which also begins to flash and vibrate. Once 'full', the ALAV shows a red light and drifts away, showing a blue light only when it is ready to feed again (ALAVs Website). In version 1.0, this feeding action was the only way in which people could interact with the robots directly, although videos do show visitors reaching out in attempts to catch hold of or hug an ALAV.

FIGURE 3.1   *The ALAVs (version 1.0)*
Source: Courtesy of Jed Berk and the Art Center College of Design.

Berk and Mitter focused on achieving two main goals in the creation of this robotic installation. First, they tried to make the robots captivating, such that they were able to attract and retain the attention of their visitors; and second, they allowed human visitors to share the same physical space with the ALAVs, so that robots and humans moved freely around one another and interacted directly. While the resemblance of the ALAVs to living or fictional beings was a factor in some people's interpretations of them, as is discussed shortly, the ALAVs' specific form and behaviours set them apart as different from anything previously experienced.

The behaviours of the ALAVs were carefully designed to support the assumption that they have feelings and can express emotions. The robots' somewhat tentative movements around humans, together with the feeding process, seem to indicate the presence of basic drives and motivations such as fear and hunger. In addition, their 'joyful' flocking dance promotes the idea that there is an emotional connection between members of the flock. The gentle movements of the ALAVs also mean that, while they are understood to be strange enough to be disconcerting, they are not normally thought of as threatening at all. For example, one visitor described them as 'kind of scary', but that same person still thinks that they are 'absolutely awesome' (ALAVs Website, Version 1.0 Full Video). Another visitor was obviously unperturbed by the robots, finding them instead to be 'vulnerable and funny at the same time', even 'lovable' (ALAVs Website, Version 1.0 Full Video).

It is also clear that some people who meet the ALAVs leave with the sense that they each have a distinct personality. In particular, a visitor describes one of the ALAVs as being 'a little dim-witted and sort of losing its mind' (ALAVs Website, Version 1.0 Full Video). The overt otherness of these robots, combined with their captivating presence, expression of individual characters and the physical proximity the installation encouraged, suggests that phenomenological accounts of communication which draw on Levinas' philosophy of the self-other encounter might be useful in theorising the ALAV-human interactions. At the same time, analysing human-ALAV interactions supports an interrogation of some of Levinas' assumptions, in particular moving beyond the case made for animals in Chapter 2, to introduce the idea that robots can also take part in meaningful encounters with humans within a Levinasian framework.

Levinasian encounters

Levinas was particularly concerned to find ways to describe how communication takes place between self and other while maintaining the absolute alterity of the other. He therefore formulated a conception of communication that was not based in commonality, or even the desire to foster increased commonality. Instead, he theorised encounters between the self and other in which they maintain an irreducible 'distance' from each other, even as they come into proximity (Levinas, 1969, pp. 47–50).

The sense of distance to which Levinas refers denotes the differences that are always present between self and other in the encounter, as opposed to indicating the retention of a physical separation. As Pinchevski explains, '[i]t is precisely in the irreconcilable difference of alterity', which sets the other apart from the self, 'that Levinas founds the fundamental relationship with the Other' (2005, p. 71). Thus, the 'relation to the Other qua Other is for Levinas the very beginning of, and the ultimate condition for, communication' (Pinchevski, 2005, p. 71). The conception of communication during the self-other encounter described by Levinas is therefore radically different from those suggested by traditional communication theories, where the success of communication is most often understood to be based in the commonalities that are self-evident or can easily be uncovered between communicators.

Human-ALAV interactions would seem to illustrate Levinas' conception of an encounter quite well. Visitors enter the installation space, and humans and ALAVs therefore come into close proximity as they move around one another. The ALAVs are perceived as captivating and strange, and they attract the attention of their visitors and interact with them directly in the presence of the food source. However, there is always a clear sense that an irreducible distance, understood in Levinasian terms to denote difference, remains between humans and ALAVs by virtue of the unfamiliar, clearly non-humanoid, nature of the robots. Indeed, it seems reasonable to suggest that it is the strangeness of these robots, fully appreciated as people come into close physical proximity with them in the installation space, that drives their visitors' enthusiasm and desire to observe them for a time and then interact with them more closely by 'feeding' them.

Given the radical alterity of robots such as the ALAVs it is worth considering how Blanchot extends Levinas' thought to offer an account of the self-other encounter within which difference is even more decisively embedded. In *The Infinite Conversation*, Blanchot defines three different ways in which the relation between self and other can be understood (1993, pp. 66–68). As Pinchevski clarifies, Blanchot's first relation is one which perceives the other 'as a function of the Same', to be subsumed into the same structure (2005, p. 97). This understanding of communication flows from the cybernetic-semiotic and the sociopsychological traditions' insistence that the other's communication should work within a particular process and coding structure. Blanchot's second relation is described by Pinchevski as a relation of 'coincidence and participation', relying on a fusion of the self and other into a unity that 'transcends both', thus relating to the sociocultural tradition's argument that the other must become a part of the prevailing culture and society (2005, p. 97).

The relation that develops between humans and ALAVs is not open to an easy appraisal from either of these perspectives. There is some level of structured exchange, for example, in the coloured lights used as feeding signals and the movement away from the food source once the process is finished. In addition, people and ALAVs are also brought together in the particular social and cultural context offered by the art installation, but this context is nonetheless unusual and does not follow the expected norms that are accepted outside the installation space. Though these structural and contextual aspects support human-ALAV communication, from the comments people make and their behaviour shown in videos, it would seem to be the overtly presented differences between humans and ALAVs that are most important in driving interactions between them. For this reason, a consideration of what occurs when humans visit these robots is more easily understood in terms of what Blanchot describes as the 'relation of the third kind' for which maintaining the difference between those taking part is the essential characteristic (1993, p. 68). Blanchot clarifies that 'what "founds" this third relation ... [is] the *strangeness* between us' (1993, p. 68, italics in the original). However, in contrast with Levinas, for Blanchot 'it will not suffice to characterize' this strangeness 'as a separation or even as a distance'; instead, it should be thought of as '[a]n interruption escaping all measure' (1993, p. 68). This sense of interruption – which Blanchot describes as 'an *interruption of being*' (1993, p. 77, italics in the original) – would seem to describe the relation that forms between humans and ALAVs very accurately, dependent as it is upon the insurmountable differences between them.

Levinas describes the encounter between self and other as 'the face to face', within which the other becomes accessible to the self by revealing a 'visage' or 'face' (1969, pp. 79–81). This terminology might seem difficult to apply to human-ALAV interactions. Although in some instances eyes do appear to have been drawn onto the robots, in general it is difficult to argue that these machines have well-defined faces or clear facial features. However, while it is sometimes assumed that the terms 'visage' or 'face' are used by Levinas to indicate the presence of physical faces, in particular human faces, there are many occasions when he uses the terms with reference to a more transcendent property of the other (Davis, 1996, p. 46). The 'face' is not simply a set of features that are seen or recognised; instead, it encompasses all of the ways that the other is revealed beyond any conception of the other that the self has already formed. Within 'the face to face' the other's integrity and alterity are therefore always retained; the other can never be completely comprehended (Levinas, 1969, pp. 50–51).

Levinas also clarifies that the presence of a physical face may not be required at all, since 'the whole body – a hand or curve of the shoulder – can express as the face' (1969, p. 262). The conception of the Levinasian face therefore encapsulates '[t]he way in which the other presents himself, exceeding *the idea of the other in me*' (Levinas, 1969, p. 50, italics in the original), whether that presentation involves the use of language, facial expressions or bodily movements. Importantly, as Christopher Diem notes, this 'ethical relation does not presuppose the capacity for linguistic articulation' (Diehm, 2000, p. 53). The face of the other 'does not have to speak' in order to make its appeal for recognition, '[w]hich means that the non-human could not be ruled out on the basis of an inability to converse with us' (Diehm, 2000, p. 53). However, in spite of the potential for expressive robots such as ALAVs to reveal a 'face' through their behaviours, the question of whether robots can take part in 'face to face' encounters is nonetheless in question.

As was briefly mentioned in Chapter 2, Levinas argues that an animal cannot reveal a face 'in its purest form' (Wright et al., 1988, p. 169), making it reasonable to assume that he would have also have excluded machines, even robots designed specifically to interact with humans, with a similar argument. At one point in his writing he does question whether 'things have a face', noting that art might be regarded as an 'activity that lends faces to things' and finally asking: 'Does not the façade of a house regard us?' (Levinas, 1989a, p. 128). However, though he does not come to a firm conclusion, suggesting that '[t]he analysis thus far does not suffice for an answer', he is clearly sceptical and wonders 'if the impersonal march of rhythm does not substitute itself in art ... for the face' (Levinas, 1989a, p. 128).

For this reason, while Levinas' defence of absolute otherness in communication would seem valuable in considering the possibilities of human communication with nonhuman others, including animals and machines, his philosophy cannot simply be employed to do this, but must instead be purposefully extended. Both Derrida and Clark, as mentioned in Chapter 2, suggest reasons to make this extension to include animals as beings that reveal Levinasian faces, able to address the human self and take part in 'face to face' encounters. David Gunkel (2010) has also recently raised the question of whether machines might be regarded as Levinasian others from a philosophical perspective, his concern being related to the question of whether machines can, or might in the future, be moral agents. In contrast with Gunkel's aim, this book focuses on extending Levinas' theory specifically in relation to robots and their communication. My aim is to analyse how humans perceive and interpret particular robots and their behaviours, employing a Levinasian perspective to help explain people's recognition of robots as communicative individuals, worthy of attention, alongside the understanding that they are also clearly strange nonhuman others.

Entertaining the possibility of extending Levinas' ideas about communication to robots, and recognising that the Levinasian 'face' does not need to be a physical face, the pertinent question is therefore whether ALAVs are expressive in ways that reveal themselves as others, such that they are able to command the attention and response of their visitors. Berk and Mitter indicate that the success of the original ALAVs project relied on its 'ability to captivate a wide audience' (ALAVs Website). As visitors watch the ALAVs, they interpret the robots' behaviours, expressed as they explore, dance and call to one another. The feeding process then allows humans to interact directly with the ALAVs as more than observers, encouraging them to take an active part in the installation and bringing the seemingly nervous robots into close proximity with their visitors. Here the lights displayed by the ALAVs help to support an understanding of the somewhat unusual feeding action. The success these robots show in attracting and retaining their visitors' attention, drawing them into direct interactions, begins to suggest that they might be encountered as Levinasian others by visitors; however, a better understanding of their expressive capabilities is made possible by considering their nonverbal communication in more detail.

Formalising the idea of communication as more than language

In line with the analysis of human-animal relations in the previous chapter, interactions between ALAVs and with human visitors suggest that rather than just trying to do 'things with words', it is vital to accept that communication of nonverbal as well as verbal signs is valuable (Peters, 1999, p. 21). While the nonverbal communication of animals is relatively well understood by at least some humans, in particular those who work closely with dogs, the strangeness of the ALAVs provokes the adoption of a number of different levels of interpretation.

The ALAVs cannot speak, but they are read as communicative in ways that bring to mind Goffman's concept of 'small behaviors', since much of their communication flows continuously as they explore the installation space as opposed to being composed of intentional attempts to communicate pieces of information. However, Goffman's focus, on 'glances, gestures, positionings, and verbal statements' revealed by humans in interaction, is somewhat difficult to map onto ALAV behaviours, in particular their use of nonlinguistic sounds and light signals. To formalise how behavioural expression, or embodied communication, works for robots such as ALAVs, beyond the use of spoken language, it is helpful to draw on Fernando Poyatos' conception of the 'triple audiovisual reality' of communication, 'what we say, how we say it, and how we move what we say' (1997, p. 249). For Poyatos, a researcher into the subtleties of simultaneous translation, communication consists of: verbal language, speech itself; paralanguage, nonverbal voice qualities, modifiers and sounds used to support meaning; and kinesics, the body language of face, eye and hand movements, and also overall body movements, postures and manners (1983, pp. 175–178; 1997, p. 249).

Within Poyatos' framework, the ALAVs communicate using their calls as a form of paralanguage and the kinesics of whole body movements. Although the spinning greeting dance and the hovering and drifting away movement when 'feeding' are the most clearly defined kinesic communications these robots make, the consistent direction in which they move, bulbous end first, identifies a robot's front from its back and may be read as indicating where the attention of an ALAV is directed. The ALAV's choice of light – blue, flashing blue or red – acting as a precise signal of whether they are 'hungry', 'feeding' or 'full' might best be regarded as a form of robot-specific paralanguage or maybe a kind of gestural action akin to the kinesics of a facial expression. An acknowledgement of the importance of paralinguistic and kinesic signals, in the absence of humanlike language, offers a way to understand how these non-humanoid robots are able to express themselves to humans, while also retaining space for their otherness to be an important part of the communicative relations that develop.

The videos of human-ALAV interactions indicate that a number of visitors to the installation do not regard the ALAVs as passive objects (ALAVs Website). Instead, the movements of the robots, the sounds and the light signals they use while interacting with one another and with humans encourage people to encounter and respond to them as expressive, thought-provokingly strange others. Human-ALAV encounters would seem to illustrate Levinas' 'provocative speculation that the ethical relation is asymmetrical: one is responsible for the Other before and beyond being reciprocated by an equivalent concern – responsibility is my affair, reciprocity is the Other's' (Pinchevski, 2005, p. 9). Visitors' careful interactions with ALAVs involve taking responsibility 'both *for* and *to* the Other: for the Other's fate, and to his or her address' (Pinchevski, 2005, p. 75). While the robots respond in their own way to the humans, it would be difficult to argue that they reciprocate on anything resembling equal terms.

Setting aside questions relating to the underlying ethical nature of human, animal and robot being, Levinas' conception of the face to face offers a way to argue for the importance of respecting otherness and difference within communication as positive elements, as opposed to problems that must be overcome. This involves recognising that animals and machines, at least those that are read as expressive, have the potential to reveal a Levinasian face in encounters with people, although they will each reveal very different faces from one another that may well require a very different response from the self.

Making space for others and their differences

The discussion above, although it refers to a more holistic understanding of communication involving the use of both verbal and nonverbal signs, is still somewhat tied to elements of a cybernetic-semiotic theory. There remains a focus on the use of signs, although in this case nonverbal signs, to transmit pieces of information in the way that many cultures, although not all, understand a shake of the head to mean 'no'. Therefore, by considering nonverbal communication in the terms suggested by Poyatos, a space is made, even within cybernetic-semiotic theory, for the existence of communicative others that are overtly different from humans. I have already noted that Levinas and scholars who draw upon his work, such as Pinchevski, describe the cybernetic and semiotic traditions of communication as defining processes that depend on the presence of commonality and sameness for successful information exchange. From the perspective of these scholars, such models can therefore be understood as violent to the other. However, attending to the complexity of nonverbal signs and taking them seriously as a meaningful part of communication processes begins to offer a way in which even a cybernetic-semiotic model can be understood to allow some space for the overtly different other to communicate, without being forced into a position of sameness.

The ALAVs also appear to express various emotions through a combination of paralanguage and kinesics. Maybe most noticeable are their anxious 'shouts' when isolated and their joyful dances when they are reunited. These nonverbal communications support the various ways in which visitors interpret these robots. In particular, the care that these robots seem to express towards one another, taking part in joyful dances when an isolated robot returns to the flock, for example, encourages people to understand them as feeling others, as opposed to unfeeling objects. The range of expression available in the nonverbal communication of even a relatively simply non-humanoid robot such as an ALAV therefore also extends the possibilities of sociopsychological and sociocultural conceptions of communication, which have less to do with information transmission and more to do with either influencing others or sharing reactions to, and understandings of, situations in the surrounding world. People's interpretations of the ALAVs' communicative capabilities support the idea that they make attempts to connect with other ALAVs and with human visitors. In part, what drives these interpretations, in particular relating to the expressive and communicative nature of the ALAVs, is a form of anthropomorphic and zoomorphic response; although, as I clarify below, these responses are also constantly in question, since these robots have not been designed to be like humans or like animals in the sense that the robots discussed in Chapter 1 and at the end of Chapter 2 have been.

As they attempt to understand the ALAVs, some visitors clearly zoomorphise the robots; for example, one chose to describe these robots as 'virtual, non sea whales' (ALAVs Website, Version 1.0 Full Video). This particular response was probably supported not just by the flock-like, equally well understood as pod-like, nature of the installation, but also the whale-like sounds being made and the dances that take place when the robots draw close together. Others have noted a more anthropomorphic response; as I have already mentioned one of the ALAVs is seen as 'a little dim-witted', but more generally people are aware that the blimps have been given names and are encouraged to read each as having its own personality (ALAVs Website, Version 1.0 Full Video). Finally, people have recognised links with science fiction, one saying, 'all my SF books try to get [this] kind of feeling', indicating that their experiences reading fiction might inform their real-life encounter with these robots (ALAVs Website, Version 1.0 Full Video). What can be drawn out from this type of comment is that humans encountering the ALAVs attempt to understand them and categorise them in whatever way helps them to support their interactions with the robots.

Most importantly, while a level of familiarity is expressed through each of these interpretations, it is subject to continuous reappraisal, because these robots cannot, for example, be un-problematically categorised as 'like whales'. Even though the feeding action brings to mind the feeding of birds or butterflies in captivity, or alternatively the experience of a dolphin encounter, the actual process of feeding the ALAVs, including the synchronised flashing of lights, and colour-coded message as the light turns from blue to red once the robot is 'replete', clearly identifies this feeding action as unique and different from any other that humans may have experienced before. In addition, this analysis of non-humanoid robots moves the discussion away from the idea that a robot's communication with people is in part reliant on the use of humanlike facial expressions, as seen in Kismet and Jules. Instead, non-humanoid robots communicate and express in their own ways, appropriate to their specific forms and behaviours.

The Levinasian 'distance' between the ALAVs and their visitors is preserved as the robots constantly raise questions that unsettle any attempt to understand them fully through the descriptions of science fiction, zoomorphic or anthropomorphic responses. Interactions between humans and ALAVs therefore open up the possibility of what might be termed a *tempered* approach, which allows the attribution of human or animal traits to non-humanoid others in order to support partial understandings, while insisting on the importance of remaining constantly aware of the level of difference that nonetheless remains. The production of a tempered anthropomorphic or zoomorphic response offers a means to understand how communication occurs, in situations where the difference between communicators remains a valuable presence.

The ALAVs in version 2.0

The ALAVs in version 2.0 are similar to those in version 1.0, only the flock is larger and the bodies of the robots are transparent, so that their signal lights cause their bodies to glow in the now more often darkened installation space. The nonverbal behaviours of the version 1.0 ALAVs were carried into version 2.0 of the installation, but a new way to interact with the ALAVs using mobile telephones was also introduced. Visitors can telephone the ALAVs' number to hear a spoken message asking them to declare whether they are 'friend' or 'foe', and then to indicate if they would prefer to interact with the 'group' as a whole or with an 'individual' member (ALAVs Website). The behaviour of the ALAVs alters depending on responses to further questions and in some cases the movement of the human. The flock of version 2.0 robots are unnamed, with the exception of 'Odd Ball', a special ALAV that takes part in one-to-one interactions with human visitors when the 'individual' option has been chosen (ALAVs Website). The dialogues between humans and ALAVs are under very tight control, revolving around single word answers (e.g., 'friend', 'individual') to recorded questions, to work within the parameters offered by the voice recognition system being used. The interaction detailed below is an example, based on the flow diagrams on the ALAVs website, of a conversation with Odd Ball having chosen the options 'friend' and 'individual':

ODD BALL: Howdy. I'm the one pulsing bright green. They call me Odd BALL. Don't be shy, come closer.
[ODD BALL: Wait 15 seconds]
    Can you tell me how tall you are in feet and inches? For example, say five seven.
PERSON: Five feet and six inches.
[ODD BALL: a. Adjust height level appropriately, b. Log height data for courtship behaviour]
ODD BALL: How old are you? For example, say thirty.
PERSON: Thirty-two.

[ODD BALL: Log age data for courtship behaviour]
ODD BALL: I think you are female, is this correct? Answer yes or no.
PERSON: Yes.
[ODD BALL: Log gender data to courtship behaviour]
ODD BALL: From where I'm from, this is our traditional courtship dance. I've heard I'm quite the stud. Thank you for helping me better understand how we can live together in a shared habitat. I hope we meet again. Goodbye.
[ODD BALL: a. Perform courtship dance, b. Return to autonomous movement]

The aims for version 2.0 of the ALAV installation were slightly different from those for version 1.0. Berk and Mitter explain that the plan was always to increase the size of the flock in the new version. In addition, they were also interested in enhancing the level of interaction between people and the ALAVs, and the use of mobile telephones as an additional interface was introduced to reinforce the idea of the ALAVs as 'networked objects that communicate the concept of connectivity among people, objects, and the environment' (ALAVs Website). It was hoped that by having spoken conversations with the robots people would feel that they had built closer relations with them. In order to analyse encounters with the ALAVs in version 2.0 Levinas' 'face to face' can again be considered, but this time to draw on the distinction that he makes between the 'Said' and the 'Saying', and its relation to ideas of interruption.

Levinas' definition of the Said can broadly be related to the idea of the message in traditional communication theory, since it consists of the ideas, information or knowledge that an interlocutor is trying to convey in language (Pinchevski, 2005, p. 10). In contrast, the Saying can be regarded as an amalgamation of what is Said with the way that it is said, to whom it is said, as well as the broader temporal, spatial, cultural and historical setting for the moment of exchange. As Levinas explains, 'Saying states and thematizes the said, but signifies it to the other ... with a signification that has to be distinguished from that borne by words in the said' (1980, p. 46). Pinchevski clarifies that for Levinas, 'communication is irreducible to the circulation of information' because it also involves 'an unrepresentable relation, contact or touch' (2005, p. 11). Therefore, while it might be argued that the Said could also encompass the distinctive signs used in nonverbal communication, it also seems reasonable to consider that much of what is conveyed by nonverbal means constitutes the very 'relation, contact or touch' of the Saying to which Pinchevski refers above (2005, p. 11).

In the case of the ALAVs in version 1.0, it could be suggested that their communication consists of precisely those elements of the Saying that are not a part of the Said. Indeed, there are only a few aspects of the communication between humans and ALAVs that result in information being conveyed, an example being the red light indicating the end of the feeding process. Pinchevski's discussion of the Saying and the Said can therefore be understood to emphasise the importance of the body and its presence in the Saying. However, analysing the new interactive possibilities of version 2.0 enables a consideration of the most important aspect of the Saying: the possibility for interruption, where this is different from the interruption in being that has already been discussed.

For Levinas, as well as for Pinchevski, a vital element of the Saying is that '[t]his signification to the other occurs in proximity' (Levinas, 1980, p. 46). While this does mean that the Saying is closely linked with the concept of 'the face to face' encounter and the interruption in being that it retains, a simpler understanding of the possibility for interruption is also present in the Saying. This occurs, as Blanchot notes, because the discourse of self and other 'is composed of sequences that are interrupted when the conversation moves from partner to partner' (Blanchot, 1993, p. 75). In this way, Blanchot links interruption with turn-taking in dialogue. Extending this idea, a related but more potent form of interruption in dialogue is actually revealed when an interlocutor simply cannot wait for his/her turn.

Although people talk to the ALAVs in version 2.0, the level of scripting in exchanges with the robots, even with Odd Ball as an individual, would seem to reduce the potential for either form of interruption in dialogue. In some ways, the Saying that occurs between humans and ALAVs is so confined through its scripted nature that it becomes more easily associated with Levinas' conception of the Said. This interaction in speech therefore does not really add to the encounter between human and ALAV in Levinasian terms. Instead, it might even be understood to reduce the impact of their nonverbal paralinguistic and kinesic expression, thus emphasising the ways in which their communication can be compared with, and found deficient against, a human standard. Videos on the website show that individual people's interactions with the robots were longer and more complex in version 2.0, but it is difficult to tell whether the robots, Odd Ball in particular, were perceived as more engaging than version 1.0 ALAVs.

Encounters between humans and ALAVs illustrate the possibility of extending Levinas' 'face to face' to include nonhuman others that reveal absolute alterity most clearly through nonverbal means. Although human-ALAV interactions offer a useful illustration of the interruption in being produced by overt otherness in communication, the communication between humans and ALAVs does not develop very far past the moment of the initial encounter. The longer interactions developed through the use of a mobile phone interface, and the introduction of communication in human language, do engage people and ALAVs over extended periods of time, but the otherness in these encounters might be obscured by the scripted nature of the conversation. The next chapter therefore considers another art installation within which communication also uses the kinesic channel, alongside a very different use of human language that may be more effective in ensuring that the differences of the other remain in mind.

4
Stories and Dances

Abstract: *In Chapter 4, the discussion moves beyond the initial encounter, to consider how dynamic interactions support communication with robots where those communications are also framed by a backstory. The focus is on how interactions can be understood in terms of both dialogue and overlapping continuous systems of interchange. Levinas' theory is further extended in this chapter, to highlight the interruptions in being and in saying that occur in interactions with Fish and Bird, the wheelchair-like robots discussed throughout the chapter.*

Sandry, Eleanor. *Robots and Communication.* Basingstoke: Palgrave Macmillan, 2015. DOI: 10.1057/9781137468376.0010.

The previous chapter explored ideas about communication in brief interactions with non-humanoid robots in an art installation. The encounters between humans and ALAVs were considered in terms of Levinas' 'face to face' with the suggestion that the radical alterity of these robots, revealed through their behaviours, including movement, sound and the display of coloured lights, can be interpreted as a form of nonhuman Levinasian face. This chapter moves on to consider human interactions with robots that also illustrate the idea of the encounter, but where people's understanding of the robots is framed by a backstory. In this example, visitors are able to explore the existing relation between a pair of robots, as well as interacting with the robots themselves. This has the potential to enable a deeper exploration of the self-other relation, developed through a process of dynamic communication between human and robot.

The Fish-Bird project

The Fish-Bird project was the result of a long-term collaboration between the artist Mari Velonaki and roboticists Steve Scheding, David Rye and Stefan Williams, working in the Centre for Social Robotics at the University of Sydney. This robotic art installation consists of two autonomous robots in the form of wheelchairs, which interact with each other and also any visitors who enter the installation space. The robots have individual personalities and are identifiable by their colour, fish being blue while bird is red. Fish and Bird communicate in two ways. The first is kinesic and relies on their movements, which are attuned to the positions and movements of the other robot and any human visitors. The second is through human language in the form of fragments of written text, which are produced using miniature thermal printers and dropped onto the floor. These notes may be personal messages from one robot to the other or messages for human visitors. The words are taken from donated love letters, the works of the poet Anna Akhmatova and a text written by Velonaki herself. Each robot produces its notes using a different handwriting 'assembled from digitized bitmaps of the glyphs' (Velonaki et al., 2008, p. 6). Figure 4.1 is a composite image that gives an impression both of the movement of Fish and Bird in relation to people within the installation, as well as showing how the printed notes build up on the floor around the robots.

64 Robots and Communication

FIGURE 4.1 *The Fish-Bird project*
Source: Courtesy of Mari Velonaki and the Centre for Social Robotics at the University of Sydney.

These robots have been carefully programmed to make subtle movements, some of which have also been linked into choreographed sequences as they move around each other. People's interpretations of their movements, and their relationship with each other, is framed by a backstory, Fish and Bird being characters in a Greek myth. In keeping with their ancient namesakes, the modern-day Fish and Bird have fallen in love, 'but cannot be together due to "technical" difficulties' (Velonaki et al., 2008, p. 5). Overall, the robots have seven patterns of behaviour, linked to the seven days of the week, and these affect the development of 'artificial "emotional" states' on a day-by-day basis. Their emotional states, or 'moods', are also shaped as responses to their perceptions of the movements of the other robot and of humans who enter the installation. Each robot's emotional state not only informs its movement, but also alters the choice of text for printing (Fish-Bird Project Website).

Interruptions in being and in saying

When they are together without visitors, the robots take part in a constant stream of interaction with each other, indicated both by their movements and by their choice of printed texts. The continuous motion of Fish and Bird, in the absence of any obvious propulsion system, complicates the idea that these objects might be everyday wheelchairs, in spite of their appearance. When a visitor enters their installation space the interaction between Fish and Bird is interrupted and the robots turn towards the person who has entered. Visitors can therefore be understood to 'disturb the intimacy of the two characters' (Velonaki and Rye, 2010, p. 3). The movement of these robots clearly shows their acknowledgement and response to the human's arrival, further destabilising their status as wheelchairs.

The Fish-Bird project demonstrates, more clearly than the example of the ALAVs, the sense that a moment of encounter and interruption can be understood to occur both for humans, as the robots turn to 'face' them when they enter the installation space, and for the robots, as human visitors disturb their conversation. The people visiting are interrupted, in the sense of experiencing Blanchot's *'interruption of being'* (1993, p. 77), because of the way that these seemingly familiar objects act in a most unfamiliar way as they turn to 'face' the person. In contrast, Fish and Bird break off from their dance and production of notes, which together could be regarded as forming a somewhat disjointed dialogue. A human entering the installation can therefore be understood to interrupt the robots' ongoing conversation.

Fish and Bird's communication complicates and blurs Levinas' ideas about the Saying and the Said because of the way in which their Saying is produced in writing, and thus closely linked with Levinas' conception of the Said. While the notes themselves cannot be interrupted during their production, interruption is nonetheless clearly present here as each robot's choice of note is altered based on what is happening around them. As I have already mentioned, this is particularly evident when Fish and Bird cease to exchange personal messages as a person enters the installation space. What they do next is begin to talk about the weather. These robots can be understood to use a similar strategy to people in the presence of a stranger, falling back on trivial topics as a means of dissipating what might otherwise be an uncomfortable silence. As the notes fall to the floor, the Saying of the robots can be understood to change into what has been Said. This record of their 'speech' to each other is always fragmentary, because no history is formed of the narrative structure of the interchange between the robots. However, since the production of notes allows visitors to see the messages that have been produced before they arrive, people may still become aware that their presence has not only altered the movement of the robots but also the content of the texts they produce.

The clarity with which the robots Fish and Bird can be understood to turn to 'look' at entering visitors returns this discussion to the idea of the face and the gaze of the other. These robots do not have eyes; information about the presence and movement of people is captured by cameras at the corners of the installation space and wirelessly transmitted to the robots. However, their form offers a clear idea of the direction in which they are facing and thus, as their attention transfers to a visitor, they are interpreted as 'looking' by virtue of their orientation. Fish and Bird therefore, more clearly than the ALAVs, lend themselves to analysis in light of Derrida's consideration of the animal gaze in his encounter with a specific 'little cat' (2002, p. 374) discussed in Chapter 2.

When entering the Fish-Bird project installation space visitors experience the sense that they intersect their gaze with the perceived 'gaze' of the robots, as Fish and Bird turn to face them. Although Fish and Bird cannot look at their visitors in anything resembling the same way that Derrida's cat gazes at him, their movement is enough to suggest that their attention is directed towards their visitors. In turning to face people Fish and Bird appear to 'look at them ... and in a word, without a word, address them', and it is this action which would seem to set the scene for further interactions between robots and humans (Derrida, 2002, p. 382). Velonaki and Rye describe this 'first stage of engagement' between the humans and the robots as a 'state of "Interest" ' (2010, p. 5). By interpreting the initial meeting of robots and humans using the frame offered by Derrida, this first stage of engagement can be understood more precisely as the moment in which visitors recognise the robots as revealing an 'absolute alterity', while at the same time retaining the ability to communicate with humans from this position of difference (Derrida, 2002, p. 380).

Interactions with Fish and Bird were designed not only to be 'intuitive, and nonthreatening' but also to be in some way 'natural', this being linked with the idea of 'the machine having a physically embodied "persona" ' (Velonaki et al., 2008, p. 2). Indeed, the form of these robots was chosen because of the sense in which the emptiness of the wheelchair, as an 'object that almost perfectly frames the human body', draws attention to the absence of a person (Velonaki et al., 2008, p. 5). In some ways, although not anthropomorphic in itself, their form was designed with the potential to invoke an anthropomorphic response to the missing person. However, it is interesting to note that in all five countries where the robots have been exhibited visitors have 'reported that they were attracted to the robots not because of the way that they looked, but because of the way that they behaved' (Velonaki and Rye, 2010, p. 5). Thus, as Derrida suggests is the case for the cat, these robots first address the human visitors by their movement, orienting themselves such that they are face to face with people.

There would seem to be no conclusive evidence that people look at Fish and Bird and think about 'what is not present' (Velonaki et al., 2008, p. 6). Instead, visitors concentrate on attempting to interact with what *is* present: the wheelchairs themselves. In addition, while Velonaki and Rye indicate that some visitors 'tended to interpret some of the robots' actions in terms of their own prior experience with people or animals', the need to zoomorphise these robots may be less marked than with the ALAVs (Velonaki and Rye, 2010, p. 5). Fish and Bird clearly do not live in the same way as their animal namesakes; however, they would seem to be good examples of the type of 'lively' machines that Haraway notes are becoming increasingly common (1985/2000, p. 294).

As I have considered in relation to Derrida's discomfort under the gaze of a cat (2002, p. 374), and Smuts' description of interactions with dogs as social subjects (2001, p. 118), recognising individual communicators is an important part of deciding to interact with them. Fish and Bird are not only separable by the colour of their upholstery, but also by the way that they move as people enter the installation. Bird is more outgoing as 'the wheelchair that first approaches an audience member', whereas Fish 'tends to hang back and observe, and is less likely to approach a person directly' (Fish-Bird Project Website). The movements of the robots therefore offer an insight into their individual personalities or characters. In addition, even though these robots' behaviours are programmed as explained above, in a similar way to the ALAVs, each individual visitor's interaction with these robots will take its own particular course.

In addition to showing where their attention lies, the movements of Fish and Bird have also been carefully designed to show their intention and mood through the speed and direction of movement. For example, '[a] robot indicates dissatisfaction or frustration during interaction...by accelerating to a distant corner, where it remains facing the walls until its "mood" changes' (Velonaki and Rye, 2010, p. 3). Wheelchairs are normally objects that require direct intervention in order to fulfil their purpose, either being propelled by the effort of the person sitting in the chair, by someone pushing the chair or by the presence of someone controlling a motorised system. Fish and Bird, however, subvert this accepted understanding of the wheelchair as passive, because of the way that they propel themselves, and cease to move at all if they are pushed or sat upon. This may well make their seeming 'aliveness' even more arresting, and may add to the way in which their movements are understood as intentioned and meaningful.

Whereas much of the movement of the ALAVs is understood as exploratory and related to a continual search for 'food', the constant motion of Fish and Bird is presented in a different way. The Greek myth from which their names are taken positions the robots as individuals who are continuously dancing around each other in an attempt to negotiate their difficult relationship. This dance is interrupted only by the arrival of human visitors or the end of the day. The myth also frames the messages that are strewn over the floor as time passes, allowing people to put the fragments of text they see into the context of this same conversation about a personal relationship. The backstory for these robots therefore helps visitors to interpret them and their communications, providing a frame for the installation's human-robot engagements that encourages people to acknowledge each robot as a communicative individual with something to 'say'.

Dialogues and dynamic systems

The creators of Fish and Bird suggest that dialogues develop between the robots and humans as the wheelchairs move around based on their sense of the 'body language of the [human] participants', who then proceed to react in their turn to 'the body language of the robots' (Velonaki et al., 2008, p. 6). During this 'second stage of engagement', 'Exploration', visitors become active participants 'by moving with and sharing the same physical space with the wheelchairs/robots' and begin to experiment with possible ways to interact with Fish and Bird (Velonaki and Rye, 2010, p. 5). Visitors try various strategies including 'making sounds such as clapping their hands or talking with a variety of different intonations to attract the attention of the robots', although they soon discover that 'physical proximity to the wheelchairs and manner of movement, changes of body stance, hand and arm gestures' are more effective (Velonaki and Rye, 2010, p. 5).

While interactions with Fish and Bird can be understood to result in dialogues or even conversations, there are alternative explanations for their communication. In particular, it is worth focusing on the way in which the reliance on a nonverbal kinesic communication channel in the Fish-Bird project, together with the production of notes as opposed to spoken language, promotes an understanding of communication which is less about turn-taking in dialogue, and more about a continuous process in which signs overlap even as they are produced by the participants. This setting aside of the turn-taking rules that are often a feature of communication with humanoid robots, as discussed in Chapter 1, does not mean the participants are not paying attention to one another; instead, the moments of attention and response occur in a more dynamic and flowing way. Therefore, alongside the idea of dialogue, it is also clear that the 'dance metaphor', introduced in the discussion of human-animal communication in Chapter 2, is an appropriate way to explain Fish and Bird's communications, both with each other and with human visitors. Not only have these robots been programmed with 'the capability to perform detailed "choreographed" sequences' as part of their movements (Fish-Bird Project Website), but there is also a clear sense of dance present more broadly in the patterns that humans and robots make as they move around one another in the installation space.

Shanker and King, whose work discusses the dance metaphor in some detail, draw upon the work of Alan Fogel, who has developed a conception of co-regulation as a process that 'occurs whenever individuals' joint actions blend together to achieve a unique and mutually created set of social actions' (1993, p. 6). Fogel stresses that '[c]o-regulation arises as part of a continuous process of communication, not as the result of an exchange of messages borne by discrete communication signals' (1993, p. 6). It is therefore possible to consider co-regulation only from a perspective which understands communication as consisting of information moving in a 'continuous process system' – or 'dynamic system' to use Shanker and King's terminology – as opposed to a 'discrete state system' (Fogel, 1993, p. 65; Shanker and King, 2002, p. 605).

Drawing on the descriptions of communication already employed in this book, it is the cybernetic-semiotic conception of communication that is most easily placed within a discrete state system. In such a system 'there are senders and receivers' and '[t]he purpose of communication is for the sender to alter the behavior of the receiver by transmitting informative messages' (Fogel, 1993, p. 65). In contrast, as Fogel clarifies in another paper, in a dynamic system '[w]ords, gestures, and expressions can be altered in their shape, intonation, size, explicitness, duration, clarity, force, and on many other dimensions depending upon the ongoing and simultaneous flow of communicative actions' (2006, p. 13). Robots such as Fish and Bird, and the ALAVs, in their interactions amongst themselves and with human visitors, can be understood to demonstrate this type of 'ongoing and simultaneous flow of communicative actions' (Fogel, 2006, p. 13). Although they do not use '[w]ords, gestures, and expressions' in the same ways as humans (Fogel, 2006, p. 13), they nonetheless express themselves through their own forms of language, paralanguage and kinesics, to use Poyatos' terminology.

Shanker and King argue that '[t]he shift from the transmission metaphor to a dance metaphor represents...a fundamental shift in communications theory from an *information-processing* to a *dynamic systems* paradigm' (2002, p. 607, italics in the original). They therefore seem to associate information processing only with what Fogel terms a 'discrete state system' (Fogel, 1993, p. 65). However, within the responses to their paper some critics note that the dynamic systems paradigm is, in fact, compatible with information processing theory (Kuczaj, Ramos and Paulos, 2002). Indeed, the cybernetic tradition, and its characterisation of communication as information processing from sender to receiver, also suggests that '"*[m]eaning*" consists of functional relationships within dynamic information systems' (Craig, 1999, p. 134, italics in the original). This is because some theory in the cybernetic tradition not only considers the importance of feedback from receiver to sender within complex systems, but also identifies a place for the emergence of new information or behaviour within a system as a whole (Craig, 1999, p. 142). Thus, rather than setting up the dynamic systems paradigm as a replacement for the discrete state paradigm, it would seem better to regard them, as Fogel does in his more recent work, as reflecting the 'different points of view of the observers and different processes of engagement with the data: one more quantitative and the other more qualitative' (2006, p. 13).

Employing both discrete state and dynamic systems models

The idea that both models can be used to understand the same interaction in different ways is illustrated in the case of Fish and Bird, and there are instances when Velonaki's consideration of the dialogue or conversation between the robots and humans is clearly an appropriate way to understand their communications. For example, the position of sender and receiver is evident as the written texts are produced, although the meaning of the texts themselves is often cryptic. A record of the sender of each note is also retained even after the note has been dropped on the floor, because of the unique 'handwriting' used by each robot. In addition, some movements, such as the initial turning of the robots towards the human visitor who enters their space, can be understood as a form of message demonstrating the robots' shift in attention onto the person. In these situations it can be argued that information is being transmitted from a sender to a receiver; the information, its source and the recipient are overtly present, although the information may not always be decoded with ease.

However, it is also clear that the movements of the robots around one another, and around humans, as well as the history of messages on the floor, promote a different idea of communication in the installation space as a constant flow of continually altering co-regulated expression. Understanding communication within a dynamic system such as this 'requires a completely different conceptualization of information that is not fixed in advance and not "transmitted"' (Fogel, 2006, p. 14). Instead, from this perspective, 'information is created in the process of communication', and 'meaning making' becomes an emergent outcome of the 'process of engagement' between humans and robots (Fogel, 2006, p. 14).

What is particularly important about dynamic systems approaches to understanding communication is that '[r]esearch on the dynamics of meaning-making admits to an inability to know completely (since behavior is changing in the very act of observing and conceptualizing)' (Fogel, 2006, p. 11). This sets it apart from traditional communication theory, and the assumption that to know the other and the other's behaviour completely is the ultimate goal of successful communication. As Fogel notes, the idea of attaining complete knowledge of the other is related to Western philosophical ideas that Levinas describes as 'totalizing' (Fogel, 2006, p. 11 referring to Levinas, 1969). In contrast, the admission of dynamic systems research, that such complete knowledge of the other is forever elusive, can be linked to various phenomenological theories of communication, including those drawn from Levinas' ethical philosophy. Dynamic systems models, and their descriptions of inter-individual communication, are open to the idea that such communication will always be 'infinitizing', with the result that one cannot 'completely know another person or completely describe behavior' (Fogel, 2006, p. 11 again referring to Levinas, 1969).

Whether communication between humans, Fish and Bird is considered as a form of dialogue or a dynamic system of overlapping communications based on body movement, if it results in the robots becoming 'comfortable' with the visitors then they may begin to reveal their more intimate thoughts in printed messages once more. At this point, in addition to writing notes for each other, they begin to write notes for people. This results in a 'third stage of engagement', 'Emotional Involvement', as visitors are drawn in by the notes that may even contain 'requests to set the wheelchairs free' (Velonaki and Rye, 2010, p. 5). Survey results indicate that 160 out of the 163 people who responded 'stated that they felt empathy for Fish and Bird caused by the messages that they received from the robots', all of them choosing 'to take their messages with them when leaving the installation space, as a memento of their encounter with Fish and Bird' (Velonaki and Rye, 2010, p. 5).

The idea of producing simultaneous and overlapping messages during communication would seem very different from the idea of taking turns, with which communication is more often associated. Indeed, much of the discussion about communication in this book relies, sometimes to a lesser and sometimes to a greater extent, on the idea of a structured development of dialogue between interlocutors. The robots Data and Kismet in particular are described as careful to adhere to turn-taking rules in conversation with humans, and this can be understood to support the sense in which they are 'polite' in their attendance to others. While Kismet's design stresses that the use of turn-taking rules acts as a means of embedding this robot in a particular sociocultural context, more generally the idea of a structured dialogue is most clearly linked with the cybernetic-semiotic tradition and the use of language.

In contrast, the areas of research which Shanker and King identify as using the dance metaphor, ranging from research in ape communication to research into infant development, are examples for which nonverbal communication is seen as a key part of the communication taking place. As has been illustrated above, with the brief consideration of Fish and Bird's communication from this perspective, it seems that the dance metaphor, and communication in a dynamic system that it describes, can also be useful in considering human-robot communication. This idea continues to be key in analysing interactions between humans and another non-humanoid robot in the next chapter. Moving away from art installations and back to a technology laboratory, the idea of fluent interchange, overlapping and flexible as opposed to following precise turn-taking rules, is embraced as a way to help humans and a robot work together to complete a task. Instead of the communication of the robot in the next chapter being framed by a backstory, this time the human-robot interaction is positioned in terms of the joint task, which the human-robot team is asked to complete. In this situation it is the task itself, and the shared experience of learning about how to complete that task as a team, which acts as a frame to support communication between humans and this nonhuman other.

5
Collaboration and Trust

Abstract: *Chapter 5 considers what happens when humans and robots learn to complete tasks together as a team. In particular, it discusses human interactions with AUR, the robotic desk lamp. In this example, elements of verbal and nonverbal communication are combined in a dynamic communication that involves paying attention to each other as well as to the task at hand. The chapter considers communication with AUR in terms of a companion species relation, drawing on the discussion of human-dog agility teams in Chapter 2.*

Sandry, Eleanor. *Robots and Communication*. Basingstoke: Palgrave Macmillan, 2015. DOI: 10.1057/9781137468376.0011.

The discussion of communication in the previous chapter is mainly concerned with how an understanding of communication as a dynamic system helps to explain human interactions with Fish and Bird. These interactions, based on and framed by stories and histories, are developed through a dance of interaction. The communication between humans and these robots reflects Smuts' suggestion that embodied communication, such as that she experiences with her dogs, is primarily about the relations themselves, as opposed to concerning 'something "out there" in the world' (2008b, p. 137). While they offer various new ways of envisioning human-robot communication, interactions with Fish and Bird do not illustrate the possibilities of collaborations between humans and robots as they work together to complete a joint task.

However, just as theorists such as Haraway and Hearne, discussed in Chapter 2, argue that working interactions between humans and animals can be of great value, this chapter moves on to consider whether similar working relationships can develop between humans and robots. In particular, this involves extending Haraway's concept of a 'companion species' relation towards an analysis of human interactions with non-humanoid robots. While the names Fish and Bird link these wheelchairs with their animal counterparts, and the ALAVs have been called a 'transitional species' (ALAVs Website), this naming strategy is really a part of their positioning as works of art: these robots are not actually like animals in many ways. It is therefore helpful that Haraway herself argues that 'the ontology of companion species makes room for odd bedfellows' and overtly includes 'machines' in her list (2004, p. 307). In addition, the previous chapter's consideration of human-robot interactions in terms of the dynamic systems paradigm and the dance metaphor is compatible with Haraway's contention that '[c]ompanion species take shape in interaction' (2004, p. 307). Unlike humans and dogs, humans and robots cannot be said to have co-evolved over a long period of time. Nevertheless, while they have no evolutionary basis, human-robot relations can still be considered as relatively well developed, since the robots under discussion are specifically designed to take part in interactions with humans.

The following consideration of the robotic desk lamp AUR involves an analysis of its position in a companion species relation, with human participants asked to collaborate with the lamp in completing a joint task. This analysis suggests that interactions between humans and robots can develop into working relationships. However, it is vital to move away from Smuts' suggestion that embodied communication refers only to the relation itself, in order to consider the ways in which embodied communication is also able to refer to external objects and processes. By supporting this move, companion species theory offers a means of transferring the ideas about human-robot communication discussed in the previous chapters into a working environment.

AUR the robotic desk lamp

AUR, sometimes described as a 'collaborative lighting assistant', was designed and built at the MIT Media Lab by Guy Hoffman (AUR Website) (Figure 5.1). AUR was developed as a means of investigating ways to improve on human-robot interactions that are typically 'unintuitive, restrictive and limited to a rigid command-and-response structure' (Hoffman, 2007, p. 23). Hoffman's aim was to find ways to support interactions that showed a *'fluency of joint action'* similar to that shown in many human-human interactions, and indeed in the human-dog interactions of agility teams discussed in Chapter 2 (2007, p. 23, italics

FIGURE 5.1   *AUR*
Source: Courtesy of Sam Ogden, photographer.

in the original). Hoffman argues that robots must 'display a significantly more fluent meshing of their actions with ours, if they are truly to enter our daily lives in a socially meaningful manner' (2007, p. 23). In contrast with his thesis supervisor, Breazeal, whose robot Kismet was discussed in Chapter 1, what is notable about Hoffman's work is that he does not assume that achieving fluency in interaction is reliant on a robot's humanlike form. Instead, he is committed to building robots that embed an autonomous ability to carry out tasks with humans into familiar workplace tools, such as desk lamps; hence the development of AUR.

The robotic desk lamp, AUR, responds to vocal commands as well as human movements and gestures (enabled by the human's use of a special glove and headband to help the robot easily locate the hand, head and head orientation of the human participant). When a human and AUR are set a task to complete, the robot is able to learn, at the same time as the human learns, what movements and light colours are required. Therefore, as the experiment proceeds, AUR is increasingly able to anticipate the human's commands, allowing it to act appropriately with less and less human direction. AUR was designed with two particular conceptions in mind to drive its ability to support 'fluent joint action' (Hoffman, 2007, p. 25). The first of these was the idea that a level of anticipation, both about 'world states' and about 'the actions of a collaboration partner', is vital in supporting effective interaction between individuals who have been asked to carry out a task together (Hoffman, 2007, pp. 24–25). The second was the idea that 'repetition, practice and rehearsal' is a valuable technique to improve the fluency of a team in carrying out a task (Hoffman, 2007, p. 25). In light of the previous two chapters' consideration of nonverbal aspects of communication it is also particularly pertinent that one of Hoffman's aims with AUR was to consider the ways in which 'co-located partners' nonverbal behaviour' is of particular importance in enabling 'joint action fluency' (Hoffman, 2007, p. 26).

Collaborating with AUR

A set of experiments, in which AUR and a human participant collaborated together to perform a simple task, was carried out at the MIT Media Laboratory as a means to test the ability of humans and AUR to work together effectively. There were three stations, or lecterns, and the aim was for the human to direct the robot using voice commands, body and hand movements such that AUR turned to each of the stations in the correct order. Once facing the station the robot was then asked by the human participant to change its lamp colour according to a written instruction hidden beneath one of a set of covers on each station. The robot had two different modes of operation, 'reactive' and 'fluency' (Hoffman, 2007, p. 119). In neither mode was the robot programmed with any prior information about the order of tasks.

In reactive mode, the robot followed the human's body movement, hand gestures and voice commands, relying entirely on the human participant for all instructions required to complete the series of tasks. In fluency mode, programming that allowed the robot to learn the task for itself was activated. During the fluency experiment human and robot gradually learnt the complex series of instructions together, such that they began to work better as a team. The lamp was able to anticipate what actions came next in the sequence and, therefore, fewer and fewer voice commands and movements were needed on the part of the human to complete each section of the task.

In the reactive experiment, the human moves and asks the robot to 'come here' indicating the station with a hand movement while also moving to the relevant lectern. The robot turns to face the indicated station and waits for the colour command from the human, for example, 'red' (AUR Video). There is a strong sense in which communication in this mode follows strict turn-taking rules, requiring the human to direct the robot at every stage. The success of this mode would therefore seem to rely on a cybernetic-semiotic transmission of information. The precision and clarity of the human's instructions, whether spoken or a combination of gesture and movement, and the ability of the robot to decode the instructions correctly are key to the team's success. In reactive mode, only the human is able to learn the task and therefore the efficiency gain of the team over time relies on the human beginning to ask the robot for the light colour as it is moving to the correct station, rather than waiting for it to complete the movement. Watching the human and robot complete the task in reactive mode, it becomes clear that the way the robot always waits its turn to respond with the action the human requests makes the experiment rather tedious to complete (AUR Video). The reactive experiment with AUR therefore suggests, as has been noted in other robotics research, that '[a]lthough not necessarily essential to collaboration, in order for the interaction to be natural and acceptable to humans, the robot teammates will need to exhibit some learning' (Johnson, Feltovich and Bradshaw, 2008, p. 9).

The idea of the importance of learning is illustrated by what occurs when AUR operates in fluency mode and is able to anticipate the sequence of events for itself over the course of the experiment. In this mode, the instructions of the human and the responsive actions of the robot initially follow a turn-taking style in a similar way to reactive mode. However, as the robot begins to anticipate what it is about to be asked to do, the flow of the interaction changes, becoming faster and more seamless. The process of completing the experiments with AUR in fluency mode therefore shares much common ground with dog agility competitions, although dog agility contains a long series of more complex and varied tasks. In both examples, it is the human participant who receives the instructions and is therefore responsible for directing both dog and robot in the initial stages of the trial. However, in a similar way to the training process through which dogs learn how to complete each individual task at home, AUR was able to learn the station-colour combinations as the experiment progressed.

In addition, in both dog agility and when working with AUR, the regular running of agility courses, mimicked by the completion of a number of experimental iterations, enables humans and dogs/robots to understand more accurately each other's nonverbal communication. It is the process of training and interacting over time that helps cement the 'companion species' relation between human and AUR, albeit in a simpler and less evolved form than that found between humans and dogs. In fluency mode, AUR becomes able to respond more and more quickly, to less and less obvious hand movements, as the human participant signals which lectern is next in the sequence. At the same time the human becomes better at interpreting the lamp's movements, to judge when it has already begun to turn in the correct direction and needs no further instruction.

In fluency mode the dimensions, 'shape, intonation, size, explicitness, duration, clarity, force' (Fogel, 2006, p. 13) of the human's verbal and nonverbal signals alter in response to the anticipation level of the robot. Gradually the human need not ask the robot to 'come here', or ask for a particular colour of light: the human's movements become enough of an indication in themselves (AUR Video). Watching the video of the experiment in fluency mode therefore draws attention to the 'dance' of the robot and the human, which becomes more noticeable as the experiment progresses and their movements are increasingly attuned to each other such that their communicative actions overlap. In fluency mode, AUR's communication with humans is more easily understood from a dynamic systems perspective than in terms describing either command-response structures or turn-taking dialogues.

Whether reactive or fluency mode is employed, the human's communication with AUR involves the use of voice, hand, head and overall body movement. However, AUR's communicative responses are based entirely on its movements and colour changes. This robot illustrates the suggestion that 'asking a robot to "turn left"', or in the case of AUR to 'come here', 'does not need an acknowledgement' if the requestor is in a position to 'observe the robot turning' (Johnson, Feltovich and Bradshaw, 2008, p. 10). Eliminating unnecessary communication, such as a verbal response to an instruction when simply following the instruction is acknowledgement enough, not only saves time and processing power, but also allows the interaction between human and robot to proceed fluently from one aspect of the task to the next.

Acting theory as an alternative perspective

As opposed to the communication theories employed and extended by this book, Hoffman drew his inspiration for the fluency design of AUR from theories and methodologies associated with the creative art of acting. In spite of this difference in theoretical grounding, his ideas also recognise the importance of following an 'embodied methodology' (Hoffman, 2007, p. 155). In particular, Hoffman explores the difference between Delsarte's system of acting and that of the Stanislavski method as a means to inform his robot's design and programming.

The Delsarte system is based on 'an elaborate analysis of facial and bodily gesture', identifying these specific movements as signs that could 'be understood as a language' (Maltby, 2003, p. 400). This basis for acting is a type of nonverbal semiotic process, which suggests that 'an audience does not intuit a character's emotion, but recognizes it through a process of signification' (Maltby, 2003, p. 400). Such a theory suggests that actors should be able to 'learn the appropriate pose to communicate pity, despondency, or pride', in the knowledge that 'the audience comprehends emotions by recognizing their signs' (Maltby, 2003, p. 400). It might be suggested that this idea of using a standardised system of common bodily 'signs' is particularly well illustrated in the facial expressions of Kismet. There is certainly a sense in which this robot was designed to produce archetypal basic expressions that could be easily recognised by people visiting the laboratory or taking part in experiments with the robot.

In contrast, the Stanislavski method is understood as a psychological theory of acting because it concentrates 'on the means by which an actor produces signs of emotion', and therefore on the idea that actors must think through a character's motivation based on their own experiences (Maltby, 2003, p. 400). As Hoffman notes, there are some problems with considering AUR from this perspective, since it doesn't have a large bank of its own experiences upon which to draw (2007, p. 157). However, more importantly for the design of this robot, Stanislavski's theory and training methods place an emphasis on the need for actors to be responsive to one another and to work together as an ensemble (Moore, 1960; Benedetti, 1998). In particular, Hoffman notes that such theory often stresses the importance of 'the space between the two actors', quoting Sonia Moore's insistence that actors 'must coordinate [their] behavior' and provide a 'continuous inner and external reaction to each other' (Hoffman, 2007, pp. 160–161 quoting Moore, 1968). Having decided how his or her character is motivated, each actor is then expected to respond to the other actors in the scene according to that motivation. This places an emphasis on the development of a relationship between actors on stage, as opposed to an exact reproduction of scripted words and actions.

Hoffman argues that the move to an 'embodied methodology' denotes a 'significant revolution' in acting method, which resonates with changes in cognitive science and also artificial intelligence research where roboticists such as Brooks suggest that embodiment is vital to the creation of intelligent machines (Hoffman, 2007, p. 155). I would suggest that it is not only a focus on embodiment that is important in marking the change in acting method and in artificial intelligence research, but also the importance of the body being situated, such that it is able to perceive and respond to the environment as it changes, as opposed to following a predefined script or map (Brooks, 2003, pp. 51–52). AUR's design can be seen to rely on these principles of embodiment and situatedness, since this robot responds to its human partner, looking for guidance from hand and voice signals, altering its response as the human – and in fluency mode as AUR itself – learns the sequence of colours and movements required to complete the experiment.

The ideas that emerge from Stanislavski's acting theory would seem to resonate with ideas of dynamic systems and the dance metaphor, in particular the sense in which 'meaning making' is understood to emerge from the 'process of engagement', whether this is for humans acting together on stage or film, or for humans and robots in laboratories, art galleries or other spaces (Fogel, 2006, p. 14). Links are also evident between the idea of attention and responsiveness to other actors within this acting theory and Fogel's understanding of co-regulation arising 'as part of a continuous process of communication' (Fogel, 1993, p. 6). In addition to these similarities, a consideration of acting theory emphasises the interplay between rehearsal and improvisation that can be observed in companion species relationships such as those between handlers and dogs in agility teams, and between humans and AUR. In both cases, practice or rehearsal plays an important part in learning about the task and in learning about one another's abilities and strengths, whereas the ability to improvise allows the team to adapt flexibly to changing circumstances.

While theories of acting method and theories of communication have separate paths of development, the resonance between them suggests there is similarity between the revolution Hoffman recognises in acting method and the shift in paradigm that Shanker and King suggest has occurred within some areas of communication studies. In particular, Hoffman's interest in applying what he felt was the 'embodied methodology' he associates in particular with Stanislavski, and his attendance to 'the technical physicality of behaviour and the conventions of nonverbal communication', has similarities with Smut's description of embodied communication (Hoffman, 2007, pp. 155 and 156; Smuts, 2008b, pp. 136–137).

However, I would also draw attention to the way in which Hoffman's design of AUR indicates that both Delsarte's and Stanislavski's ideas have a part to play. Hoffman explains that in designing AUR it was important that the robot supported 'readability of expressive behavior', a requirement that might be seen to be drawn from Delsarte's semiotic theory linking body movement and emotional expression (Hoffman, 2007, p. 156). In addition, Hoffman's design also allows the robot's behaviour to emerge as part of the process of balancing its response to the commands of the human participant against what it has learned about the task, an idea drawn from Stanislavski. This therefore suggests, in a similar way to Fogel (2006, p. 13) in his comparison of the dynamic systems and the discrete state systems paradigms, that these acting theories need not be set up in opposition to each other, such that Stanislavski's ideas are described as the basis for a revolutionary new acting theory. Instead, both the semiotic theory of Delsarte and the dynamic theory of Stanislavski have played a part in the design of AUR, as well as aiding peoples' understanding of the robot's communication.

As has already been mentioned, Hoffman acknowledges that it is difficult for robots to amass personal experiences and memories, suggesting that this might compromise their expressive abilities. In addition, he also recognises that, from the perspective of the Artificial Intelligence community, 'artificial agents will never be intelligent until they are able to accumulate experience and memories over a prolonged period of time' (Hoffman, 2007, p. 157). However, although AUR may not be able to develop a long-term personal history, experiments indicate that the personality of this robot, and an associated sense of its intelligence, is developed even with only a modicum of experience accumulated over the relatively short span of an experiment. As Hoffman explains, in spite of the difficulty of arguing for personality and intelligence in robots: experience shows that even the most simple combination of an underlying desire and affective state modulating the otherwise straightforward functional behaviour of the robot significantly changes the way people interact with the robot, attributing a much higher degree of 'understanding' and 'realism' to the inanimate object (Hoffman, 2007, p. 163).

Expression and 'face'

The expressiveness of AUR's movement depends to an extent on the way its body structure is read by the human participant. Although Hoffman has not created this robot with the aim of supporting an anthropomorphic or zoomorphic response, it is nonetheless possible to perceive this robot as having a face, head and neck. In this sense, AUR shares much in common with the lamps in the short animation *Luxo Jr.* (1986). Although many animations of inanimate objects have been made, most alter the objects to make them obviously alive with the addition of cartoon faces, as seen in the creations of Walt Disney.

However, in *Luxo Jr.*, the depiction of the two communicative anglepoise lamps does not require the addition of expressive faces to convey the story. The lamps have been animated in a way that is consistent with a real lamp's physical mechanical properties, although they are able to move their whole bodies around by springing into the air. They possess physical 'faces' and 'heads' by virtue of their bulbs and the surrounding lampshades, and their anglepoise 'bodies' are also incredibly expressive. The story of the short film is communicated in a variety of ways through the 'body' language and 'head' movements of the lamps. In particular, their gaze direction is very clear and easily understood through the direction of the beams of light produced by their 'faces'. This kinesic expression is also supported by paralinguistic sounds. These animated lamps are therefore excellent examples of simple mechanical objects that use kinesics and paralanguage to communicate intentions and feelings without the need for either human language or an expressive human face.

Although AUR's design is not quite as flexible as the anglepoise lamps in *Luxo Jr.*, this robot's 'head' and 'face' still draw the attention of the human participants, such that they consistently look to the lampshade, and sometimes into the shade as they work with it to complete the joint task (AUR Video). In spite of the way that AUR has a physical face and can change the colour of light it displays, it could be argued that this is actually not as expressive as the overall movement of its body and its posture. As is explained in more detail below, AUR's expressive body, in particular when its movements show anticipation for the human's instructions, allows it to reveal a level of personality during interactions with people. This robot's expressiveness is such that human participants experience AUR as an intelligent other, at least over the course of the experiments. AUR can therefore be said to reveal its own specific form of nonhuman Levinasian face.

Hoffman argues that developing a level of 'robot personality' is valuable in creating expressive robots that can communicate with humans, an insistence that can be linked back to Smut's description of her relationships with dogs, and the importance of her understanding of each dog as a unique social subject with a specific personality (Smuts, 2001, p. 118; Hoffman, 2007, p. 157). The effect of AUR's personality, together with the perception that it is an intelligent robot, is particularly well illustrated by the answers provided by participants in response to the open-ended questions in Hoffman's post-experiment survey. This anecdotal evidence provides possibly the most revealing evidence pertaining to this discussion of humans and robots becoming 'companion species' to one another through anthropomorphism, zoomorphism and an understanding of communication that is both dynamic and attentive to the importance of bodies and their expressiveness as well as language.

Trust, respect and interruption in dynamic interactions

Two sets of human participants took part in the experiments: one group worked with the robot in reactive mode and the other in fluency mode, but the people in each group were unaware that AUR had two different modes of interaction. The results of the experiments demonstrated 'a significant improvement in team performance' when the robot was operating in fluency mode (Hoffman, 2007, p. 127; Hoffman, 2008, p. 5; Hoffman and Breazeal, 2010, p. 412). In addition, people working with the lamp in fluency mode 'rated the robot's contribution to the team significantly higher' than people who worked with the lamp in reactive mode (Hoffman, 2007, p. 142; Hoffman and Breazeal, 2010, p. 419). Thus, in fluency mode, not only did the team's performance improve as the task repeated, allowing both human and AUR to learn the steps, but also the human participants were aware of the positive contribution that the robot's learning and anticipation made towards the improved team performance.

Even when it was simply reacting to the human's orders, AUR caused one participant to remark that at times it felt as if they were 'interacting with a being that was more alive... [than a] machine' (Hoffman, 2007, p. 147). However, the ability of AUR to anticipate both the task and the instructions was most often appreciated by those participants in the fluency experiment, one of whom stressed their 'sense of relief when it just did what I was about to tell it to do' (Hoffman, 2007, p. 147). Indeed, AUR's abilities clearly led some participants to zoomorphise the robot, possibly because of the use of 'short commands similar to the ones given to animals' (Hoffman, 2007, p. 148). Indeed this can be seen as a similarity between working with AUR and dog agility competitions, during which human team members use similarly short commands to direct the dogs around the course.

In contrast, some participants felt that they developed a relationship with the robot that was more clearly based in anthropomorphism. For example, one person suggested they 'were good friends' with AUR '[b]y the end of the second sequence', the sense of friendship being so strong that they 'high-fived mentally after the task was done'

(Hoffman, 2007, p. 147). Another participant would have liked to be able to have 'a robot as a teammate to perform any task' (Hoffman, 2007, p. 147), although they were also somewhat concerned by their reaction to AUR, noting some 'fear' at the ease with which they 'grew affectionate to it', which led them to wonder if they would 'ever have the need for it to have true (!) feelings' towards them (Hoffman, 2007, pp. 147–148). Although this person's level of camaraderie with AUR was higher than most, it is clear that participants did feel their responsibility towards the robot quite strongly, some going so far as to suggest that the robot was better at the task than they were, expressing regret that their mistakes might have slowed their 'teammate' down (Hoffman, 2007, p. 148).

From these comments it would seem that, while humans initially regarded AUR as an object, in fluency mode they were soon surprised and impressed by the way that the lamp remembered what it had been asked to do previously, and anticipated what it was likely to be asked to do next. By the end of the fluency experiments the human participants felt that they had formed a working partnership with the robot. Indeed, some participants even expressed feelings of 'respect', for the robot's memory and ability to anticipate new moves intelligently, and finally 'trust', because as a team their performance at the task consistently improved by working together (Hoffman, 2007, p. 147). The most telling representation of this trust is what occurs when a human participant makes a mistake and misdirects the lamp to the wrong lectern. The lamp does a double take; that is, it turns the way it 'thinks' it should go, and then turns back to 'look' at the human. The human sees the double take, expresses confusion and then makes a correction (AUR Video; Hoffman, 2007, p. 152). Through this momentary act of 'body language' AUR shares its knowledge of the situation, and the human has sufficient trust in the robot's understanding of the task to acknowledge the mistake and correct the instruction.

In Chapter 2, the story of Hearne and the dog Gunner draws attention to what happens when a human is not sure that their nonhuman partner is attending to the task at hand (Hearne, 2000, pp. 13–14). Hearne did not trust Gunner's motives as he rushed into the bushes, but was proved wrong when he returned with Colleen's stuffed toy. In the case of AUR, the human recognises his mistake quickly, possibly because of the nature of the task, but also because his recent experience of AUR's abilities supports the sense that the lamp 'understands' the requirements and therefore 'knows' what it should do next. The comments made by participants in experiments with AUR, in particular those who express the responsibility they feel towards the robot and the respect they develop for its abilities, suggest that they respond to AUR in a way that resonates with Levinas' idea of the other. AUR, unlike the ALAVs, Fish and Bird, is more able to reciprocate, at least in terms of appearing committed to completing the series of tasks at hand.

What is most important in comparing moments of interruption in human-dog and human-robot teams is that they illustrate the way in which a reliance on sameness, which is assumed to be the fundamental building block of communication in traditional theory, can be replaced by a similar reliance on achieving fluency in dynamic systems approaches, whether based on dance metaphors and dynamic communication theory or acting theory. Sometimes interruptions do occur from which there can be no recovery. Haraway, for example, describes an occasion when running a trial with her dog Cayenne when they both, simultaneously, lose contact with each other and with the course to be run (Haraway, 2008b, p. 230). However, AUR's double take, and Gunner's decision to retrieve Colleen's toy, stresses that moments of unexpectedness and uncertainty can be read as communicating more in a brief moment than is communicated throughout a longer period as part of smoothly co-ordinated joint action. There is clearly a need to recognise that even within dynamic frameworks, which may well be more open to otherness than static or discrete state systems, communication success – but this time in terms of fluency – can become such a strong goal that the idea of a break – an interruption – is once again seen only as a problem to be solved, as opposed to a moment in which something very important is communicated.

Part III
Rethinking Robots and Communication

6
Humans, Animals and Machines

Abstract: *Chapter 6 considers the implications of human communication with nonhuman others for the categories human, animal and machine. It argues that, while the boundaries between these types of being are becoming increasingly blurred, they are nevertheless still meaningful. The chapter goes on to consider ways of assigning agency to nonhuman others on the basis of their activity in situations, while also recognising the difference between human activities and nonhuman activities.*

Sandry, Eleanor. *Robots and Communication*. Basingstoke: Palgrave Macmillan, 2015. DOI: 10.1057/9781137468376.0013.

This book explores communication between humans and robots of many different forms, which have broadly been categorised as humanoid (in Chapter 1), animal-like (discussed very briefly in Chapter 2) and non-humanoid (in Chapters 3, 4 and 5). The book therefore emphasises how the figure of the robot can be understood to draw together the categories of human, animal and machine in discussions of form and behaviour that relate to analyses of communication. However, alongside this drawing together of disparate categories, my argument has also sought to emphasise the continued value that difference has in processes of communication. This chapter therefore considers more precisely how the boundaries between humans, animals and machines can be negotiated in such a way that communication can occur, while also retaining a clear sense of the absolute alterity of the other.

The value of blurred yet meaningful boundaries

Haraway has argued that 'nothing really convincingly settles the separation of human and animal', since '[l]anguage, tool use, social behaviour [and] mental events' are no longer regarded as solely present in humans (1985/2000, p. 293). The result of this blurring in boundary conditions is reflected in Chapter 2, where the interconnecting works of Smuts, Haraway and Hearne are used to explore the potential for rich communication, and therefore the development of complex collaborations, between humans and animals. Making this argument involves acknowledging that animals exhibit social behaviours, as well as accepting the idea that 'language' refers to considerably more than human language. From this perspective, communication is not only the transmission of information in words, but also dynamic exchanges of nonverbal paralanguage and kinesics, to use Poyatos' terminology. These cross-species interactions also highlight the importance of attending to both intentional and unintentional communications, including the 'small behaviors' defined in Goffman's model.

The division 'between animal-human (organism) and machine' has also been questioned by Haraway, who considers this boundary to be 'leaky', noting that while machines were once 'not self-moving, self-designing [or] autonomous', by the late twentieth century a stage has been reached where 'our machines are disturbingly lively, and we ourselves frighteningly inert' (1985/2000, pp. 293 and 294). Haraway is not concerned with defending the boundaries between humans, animals and machines, but rather in using their collapse to enable her introduction of the cyborg figure, whose very existence requires them to be compromised. She continues to make possibly her most quoted statement emphasising her appraisal that the boundaries have already been breached when she says, 'we are all chimeras, theorized and fabricated hybrids of machine and organism. In short, we are cyborgs' (Haraway, 1985/2000, p. 292).

N. Katherine Hayles also discusses the boundaries between humans and machines, arguing that we should not 'police' them and suggesting instead that there is much to be learnt from trying to understand the flow across them (Hayles, 2005, p. 242). Hayles clarifies that for her, the 'boundaries are *both* permeable and meaningful', such that humans can still be regarded as 'distinct from intelligent machines even while the two are becoming increasingly entwined' (2005, p. 242). She suggests that if, even as they are drawn together, a clear idea of the remaining difference between human and machine is retained, the development of 'a dynamic partnership between humans and intelligent machines' can be proposed (Hayles, 1999, p. 288). It is this idea, rather than a desire to dissolve human-machine boundaries further, that has been pursued most clearly in this book. From this perspective the ideas of proximity (with the possibility of interaction that it supports) and distance (with the continuous acknowledgement of difference that it requires) found in Levinas' conception of the 'face to face' are important not only when analysing interactions between specific selves and others, but also when thinking about the categories human, animal and machine, and the blurred, yet still meaningful, boundaries that exist between them.

Rethinking the divisions between categories of being can raise the question of who, or what, can be thought of as a 'person'. As I discussed in Chapter 2, Smuts argues that her appraisals of individual animals are not about attributing 'human characteristics to them', but rather relate to seeing them 'first and foremost as *persons*' in their own right (2001, p. 118). In contrast, Hearne's work steps back from the question of personhood to defend anthropomorphism as a valuable way to understand the behaviours of animals while keeping their otherness, and their particular skills and abilities, in mind (2000). In terms of robots, the question of whether machines should be afforded a level of personhood may be even more difficult to answer, as is clear from the detailed discussion in David Gunkel's book *The Machine Question* (2010). This book has not sought to answer, or even discuss, this question in any great detail. Instead, its exploration of human-robot communication has been focused on the possibilities of tempered anthropomorphic and zoomorphic responses to robots. These appraisals of robots support the idea that many different forms of robot can be understood by humans, and may therefore operate as part of collaborative human-robot teams, even as the differences between these machines and humans remain marked.

Recognising the difference of humanoid robots

In discussions relating to social robotics the propensity of people to ascribe human characteristics to nonhuman others is generally accepted as valuable (Dautenhahn, 1998, pp. 573–617; Turkle et al., 2006). For example, human interactions with Kismet are judged to be supported by the anthropomorphic responses this robot provokes (Turkle et al., 2006), and similar effects can be seen in interactions with Jules, and of course in fiction with Data. These robots, while differing from one another in the details of their appearance, are all broadly created as humanoid and can thus also be described as anthropomorphous robots.

In many ways, humanoid robots, by virtue of the design decisions of roboticists, force people to follow a particular course of interaction, one that is in keeping with the high level of anthropomorphism they provoke. These robots do not 'invite projection', the trait identified by Turkle for 'evocative objects' such as computers (2005, p. 27); rather, they are more easily thought of as 'relational artifacts', which 'demand engagement' (Turkle et al., 2006, p. 315). Indeed, it can be argued that humanoid robots make the specific demand that people should relate to them as if they were human. By considering humanoid robots as relational artefacts, the way in which they are less about evoking a range of interpretations and responses, and more about actively facilitating only one particular understanding and mode of interaction, is made clear.

Of course, the decision to control interactions with robots in this way is made because, by provoking a particular understanding of the robot as humanlike, it is hoped that interactions with it will be easier and more meaningful for more people: roboticists are attempting to reduce uncertainty and the misunderstanding it might cause. Breazeal, for example, argues that carefully framing her robot, Kismet, as a young child helps to support its interactions with people. In addition to the stylised design of the robot's form, this decision has also involved a careful consideration of Kismet's communication. Although in a few experiments human participants were asked to teach Kismet certain words, for example, their names, in general Kismet did not speak a fully formed human language. Instead, Breazeal's focus was to make Kismet produce a recognisable form of pre-linguistic speech, similar to the babbling of a small child. This decision positioned Kismet as childlike and also supported Breazeal's interest in the passing of affective content via prosodic cues during turn-taking dialogues (Breazeal, 2002b, p. 884).

The term 'prosody' refers to the rhythm, tone and stress used in the production of human speech, whereas the term 'paralanguage' is often used more broadly to refer not only to voice modifiers, but also to other sounds considered to be nonverbal communications separate from speech (Nöth, 1990, pp. 247–248). It may be unsurprising that Breazeal placed her research emphasis for Kismet on the idea of prosody, because this term reinforces Kismet's position as an infant that is only just beginning to learn to speak. In contrast, my decision to consider paralanguage allows me to acknowledge that nonhuman others are communicative in nonhumanlike, often nonverbal, ways as opposed to implicitly placing their communication as deficient when measured against that of human adults.

In spite of Breazeal's efforts, the idea of deficiency is present in some appraisals of Kismet, where people suggest that this robot lacks the complexity of character to maintain long-term interactions. For example, Peter Menzel notes, 'Cynthia's Kismet was as cute as any baby I have ever played with...but I was ready for a smarter robot after fifteen minutes' (Menzel and D'Aluiso, 2000, p. 69). Menzel's response might be linked with the way in which humanoid robots raise people's expectations above and beyond the actual communicative ability of the machine, leading to a level of disillusionment after interacting for a short time. However, this attitude also betrays a particular stance towards children as much as anything else, a perspective that sees them as less developed, less able and less interesting than adults. There are many more accepting perspectives on children that could be fruitful in analysing communication with Kismet. In particular, those that draw on Adlerian psychology, which encourages respectful interaction with children in ways that preserve their dignity as well as that of parents and teachers, might be employed. It could, however, be argued that Kismet's framing is such that the traditional perspective, involving a tendency to talk down to children, seems easier to follow than these alternative, more respectful, options.

Jutta Weber also critiques Kismet, when she notes that the design decisions of Breazeal manipulate users by modelling '(gender)stereotypical social relations, which rest on the anthropomorphisation and personification of machines as infantile and helpless creatures' (2005, p. 216). She argues that '[t]his concept of Human-Robot interaction exploits the readiness of the user to compensate for the deficiencies of machines' (Weber, 2005, p. 216). Weber would therefore probably be unsurprised that Menzel was not drawn in for long by Kismet's infant-like personality. However, her analysis is also clearly shaped by the perspective towards children that sees them as deficient in comparison with adults. I would argue that the idea of deficiency in the robot presented in Weber's critique, and more broadly in human-robot interaction research itself, is an unfortunate result of deciding to judge such machines against an idealised adult human standard. This appraisal, in a similar way to traditional communication theories, could even be regarded as violent to the other. A viable alternative would be to suggest that, instead of being *deficient*, the machine be regarded simply as *different*. However, the problem of keeping the difference of a humanoid robot in mind may make it difficult to regard it as a nonhuman other that should be respected for what it can do, as opposed to disrespected for its perceived failings.

Although some of Turkle's work has supported the development of robots such as Kismet, in her more recent writing she articulates her anxiety that human relationships might eventually be replaced by human interactions with sociable robots (Turkle, 2011). I would suggest that a similar anxiety might be directed towards the creation of animal-like robots designed to replace human interactions with animals. As is clear from the discussion of humanoid robots in Chapter 1, animal-like robots in Chapter 2 and further emphasised in the appraisals of Kismet above, the creation of robots that communicate socially in familiar ways is technically very difficult to achieve. In spite of this, the idea that reducing or even eliminating the otherness of machines will make interactions more successful seems to compel some roboticists to continue in the pursuit of this goal. Their assumption is that such robots might provide easy companionship, offering much of the sociability with none of the difficulties, whether that is the difficulty of dealing with the judgements and changing personality of another human or caring for the physical needs of a pet.

The additional problem that I see with robots such as Kismet and those that follow similar design paths is not only that they force people to interact with them as somehow humanlike, but also that they are limited by this design decision, because of the difficulty in creating an anthropomorphous robot that also uses novel ways to sense its surroundings and to communicate; instead, such a robot is limited to being like a human. Of course, an anthropocentric perspective that views humans as the pinnacle of evolutionary progress might also view being humanlike as the pinnacle of robotic development. Indeed, the creation of humanoid robots is seen by some as 'the Grail' of robotics (Menzel and D'Aluiso, 2000, p. 18).

There are some applications where the creation of a humanlike robot that is as close to indistinguishable from a human as possible is regarded as essential, such as in the creation of Sex Bots (Levy, 2007; Yeoman and Mars, 2012). This type of robot might eventually be expected to communicate exactly like a human, although some people might be less interested in what they say and more interested in what they do. In other situations, as is argued in the case of the Atlas rescue robot, humanlike form is understood to allow robots to enter and operate within compromised human-tailored workplaces, such as those found in the heavily damaged Fukushima Daiichi nuclear power station (Boston Dynamics Website; DARPA Robotics Challenge Website). However, by taking the perspective that the creation of humanoid robots is the ultimate goal of robotics, one overlooks both the difficulties of creating robots in humanlike form and the possibilities of creating robots that have very different ways to view and communicate about the world from humans.

As I have argued in Part II, paying close attention to the communication that occurs between humans and non-humanoid robots offers a new way to assess human-robot relations that may also provide some relief for Turkle's anxiety and for those that share similar reactions to the potential future of human-robot companionship. In particular, by highlighting the possibilities of regarding robots as communicative machines that have their own form of 'being', expressing their own type of 'liveliness', this book has gone some way towards suggesting that non-humanoid robots might form new kinds of relation with humans, which do not need to be framed as mirroring human-animal or human-human interactions. The companionship of such robots can therefore be figured as something new and different, not as a replacement for existing relations but rather as an addition.

Non-humanoid robots and difference

In her recent research, Haraway has also become more concerned with stressing the existence of difference over sameness in the development of relationships between humans and others (2003, p. 21). As discussed in Chapter 2, she has turned to human-animal relations for inspiration, in particular examining how humans and dogs are able to work as partners through her conception of 'companion species' relations (Haraway, 2003). For Haraway, 'the beauty of dogs' is that they are 'not about oneself' (2003, p. 11). In a companion species relationship, communication therefore occurs across 'irreducible difference', resulting in the development of a '[s]ituated partial connection' for which '[r]espect is the name of the game' (Haraway, 2003, p. 49). Haraway's arguments resonate with the Levinasian concern for the alterity of the other, and an understanding that interactions need not rely on an assimilation of the other to the same. A similar appraisal has led me to consider non-humanoid robots within the frame offered by the 'companion species' relation in association with Levinas' philosophy, to support my contention that robots need not be humanoid in order to support meaningful interactions with humans. In a literal sense, and in an echo of Haraway, it is the non-humanoid robots discussed in Part II of this book that can most easily be identified as 'not about oneself', unlike the humanoid robots discussed in Chapter 1.

While non-humanoid robots are still 'relational artifacts', in that they actively encourage engagement, these robots also facilitate many varied interpretations of their behaviours and any indirectly or directly communicative actions they make. They can therefore also be understood to operate as 'evocative objects'. What may be most important about the non-humanoid robots discussed in earlier chapters is that they are so clearly different from humans and animals. This means that while anthropomorphism and zoomorphism may provide useful ways to engage with their behaviour, the validity of these understandings is also continuously in question, because the robots are so obviously also machines. This has led me to suggest that when interpreting non-humanoid robots, peoples' anthropomorphic and zoomorphic responses are tempered by the clarity with which they are also understood as machines.

Haraway notes that Hearne, in her discussion relating to trainers and their animals, makes a strong case for anthropomorphism as a way 'to keep the humans alert to the fact that somebody is at home in the animals they work with' (Haraway, 2003, pp. 49–50). However, as Haraway goes on to emphasise, '[j]ust *who* is at home must permanently be in question', because '[t]he recognition that one cannot *know* the other or the self, but must ask in respect for all of time who and what are emerging in relationship, is the key' (2003, p. 50). This idea of not being able to '*know* the other' is strongly represented in Levinas' philosophy of 'the face to face' encounter, which describes the face as '[t]he way in which the other presents himself, exceeding *the idea of the other in me*' (Haraway, 2003, p. 50; Levinas, 1969, p. 50; italics in the originals). Levinas thus describes encounters during which '[t]he face of the Other at each moment destroys and overflows the plastic image it leaves' (1969, pp. 50–51). It should therefore be recognised that the idea of the robot, its personality and behaviour, which humans interacting with it formulate (often using processes of anthropomorphism or zoomorphism or both) is always a 'work in progress'. The human's partial knowledge of the robot other is 'permanently in question' as the robot's revelation of itself through its Levinasian face demands constant re-evaluation (Haraway, 2003, p. 50). As I have indicated, there are times when anthropomorphism and zoomorphism play a part in aiding people's interactions with the non-humanoid robots discussed in earlier chapters, but in all cases a sense of their otherness is also retained.

Activity and agency within a particular context

The discussions of the non-humanoid robots in this book have also highlighted some of the other reasons why people interpret them as communicative others. AUR is framed by the experimental conditions under which humans are asked to work with the robot. As the experiment progresses in fluency mode, people gradually become aware that AUR is learning the task with them, and is able to anticipate their next instruction. Working with the robot becomes easier and easier and, if the human member of the team makes a mistake, their sense of respect for and trust in AUR supports a decision to correct their instruction.

The characters of Fish and Bird are encountered within an installation space. As artworks they are explicitly presented to be observed, and interpretations of their activities are supported by the backstory of their failed love affair. People have the opportunity to watch these robots as they produce texts and dance together, gaining a sense of their active engagement with each other. If a human enters, the moment when Fish and Bird stop dancing and turn to face them strongly communicates their attentiveness to the visitor, encouraging people to try various ways to interact with the robots themselves. The ALAVs are also found within an installation space, which draws people's attention to their unusual form, echoing calls and apparent care for one another shown by the flocking dance that places them as social. The feeding process and the scripted exchanges (the latter available only in version 2.0) then allow people to interact with the ALAVs on closer terms.

Central to people's appraisals of these robots is the fact that they all appear to be actively engaged with their surroundings, with other robots (when they are not alone) and with their human visitors or experimental partners. Broadly, therefore, while I would not attempt to argue that the robots I've discussed in this book are capable of taking moral or ethical responsibility for themselves, they can nonetheless be interpreted as active agents that can take part in interactions with people. This idea, that activity is a key part of recognising agency, is supported by the work of François Cooren, who suggests that agency should be reconceptualised to include all nonhuman agents that contribute to situations. Cooren argues that recognising the presence of 'something that *makes a difference*, whether in terms of activity or performance' is enough to identify it as an agent (Cooren, 2010, p. 20).

Cooren's analysis considers a very broad range of agents, including 'utterances, emotions, collectives, principles' and 'rules' (2010, p. 9). Some of these agents, for example, 'principles' and 'rules', become active only when humans 'mobilize' them in communication 'by ventriloquizing them, that is, by making them say or do things' (Cooren, 2010, p. 9). However, Cooren's broad recognition that 'action and agency are not human beings' privileges' is important, because it allows one to 'decenter' the analysis of situations to note how 'people are acted upon as much as they act' (2010, p. 22).

The robots I have discussed in this book would certainly seem to make a difference through their activities and can thus be recognised as agents from the theoretical perspective offered by Cooren. Although it may be possible to argue that these machines communicate intentionally on some occasions – when Odd Ball the ALAV speaks, Fish or Bird produce a message for someone, or AUR shakes its 'head' in the form of a double take – it should be acknowledged that the meanings of many of their communicative behaviours are reliant on human interpretations. While it might be argued that robots do not need to be 'ventriloquized', because they are able to 'say or do things' for themselves, people's interpretations of their behaviours are nonetheless a key part of how communication with them proceeds.

For Cooren, and for my argument in this book, the benefit of 'acknowledging the broad intension and extension of the concepts of action and agency' is that it allows one to 'recognize all the various things that things do (with or without words)' (2010, p. 90). In addition, and this is key given my concern to recognise the alterity of the other, Cooren argues that extending the idea of agency in this way does not mean that 'the differences between its various forms' must be abandoned (2010, p. 4). From this perspective, the specificities that distinguish human and nonhuman activities, and therefore human and nonhuman agencies, should be retained.

This broad idea of what constitutes agency cements the position of robots of all forms as individuals that are capable of interacting with humans; however, the examples that have been discussed throughout much of this book also suggest that it is important to attend to the system of interaction itself. The next chapter therefore moves away from questions that are purely about individuals, their categorisation and their potential for agency, to consider how thinking about dynamic systems, and the communication that emerges in those systems as humans and robots interact over time, can support the formation of collaborative teams.

7
Communication, Individuals and Systems

Abstract: *Chapter 7 concentrates on exploring some ideas about the relationship between individuals and systems in thinking about communication. It discusses long-term interactions with robots outside of laboratories and art installations, identifying the value of respect and trust in collaborative partnerships with robots. This is developed into a consideration of how responsibility is shared across collaborative teams, even when the team members are in an asymmetrical relationship.*

Sandry, Eleanor. *Robots and Communication.* Basingstoke: Palgrave Macmillan, 2015. DOI: 10.1057/9781137468376.0014.

The consideration of interactions between humans and non-humanoid robots in this book has already gone some way to emphasise how these communicative situations are better understood when attention is paid not only to the actions of individual communicators, but also to the dynamic system of communication that is formed over the course of their interaction. This chapter first identifies some key factors that enable the development of fluent dynamic communication with AUR and then takes the opportunity to reassess Kismet's communication as part of a system.

All of the examples discussed in the book so far involve relatively short-term interactions. Within these situations the initial encounter between human and robot, together with the early process of learning how to communicate, is most important. The main aim of this chapter is therefore to discuss the formation of longer-term relations and associated interaction systems between humans and robots. This takes the book's consideration of human-robot interactions out of the laboratory or installation space, to analyse robots that are deployed with explosive ordnance disposal (EOD) teams, as well as robotic cleaners in people's homes.

Rehearsal or training and fluency

As discussed in Chapter 5, Hoffman's design of AUR uses the ideas of anticipation and rehearsal as foundational concepts, his aim being to explore how to make interactions between humans and robots more fluent. Hoffman argues that the 'unintuitive' and 'restricted' nature of many human-robot interactions is caused by following 'a rigid command-and-response structure' (2007, p. 23). Anthony Finn and Steve Scheding make a similar observation when they note that human interactions with robots are still 'usually based around some form of turn-taking behaviour, which can introduce delays and inefficiencies or even cause frustration' (2010, p. 50).

Finn and Scheding therefore emphasise the value of designing robots that are able to 'work more fluently with their human partners', suggesting that human and robot must be 'familiar with both the task and each other' in order for fluent communication to arise (2010, p. 50). In Hoffman's research, the idea of becoming familiar with the task and each other was linked with discussions of rehearsal in relation to acting. In experiments with AUR, the use of repetitive tasks mimicked the process of rehearsal, allowing Hoffman to measure the benefits of learning together for the team's fluency. For human-AUR interactions in fluency mode, the repetition of the task supports both the human's and the robot's ability to learn the sequence of moves necessary to complete the task. In addition, they also learn to read and anticipate each other's moves such that their interaction ceases to follow strict turn-taking rules but rather becomes a dynamic and overlapping system of communication. At this level, communication is less about the individual communicators and the messages they produce and more about the dynamic system of communication that develops around them.

Reconsidering Kismet as part of an interaction system

Although interactions with humanoid robots are designed to be fluent without the need for training, even these interactions would seem to improve when human and robot have time to learn how best to communicate with each other. This is the case with Kismet, for example, as is made clear in Lucy Suchman's description of her visit to the MIT Media Lab. Suchman notes that no one in the party with whom she visited the lab 'was successful in eliciting coherent or intelligent behaviors' from Kismet (2007, p. 246); however, in contrast with Weber's appraisal discussed in Chapter 6, Suchman's analysis takes a more positive direction.

Rather than focusing on Kismet's deficiency in communicating with visitors, Suchman suggests that Kismet should be reframed 'from an unreliable autonomous robot, to a collaborative achievement made possible through very particular, reiteratively developed and refined performances' (2007, p. 246). Her argument focuses on recognising that 'Kismet's affect is an effect not simply of the device itself but of Breazeal's trained reading of Kismet's actions and her extended history of labors with the machine' (Suchman, 2007, p. 246). Therefore, while discussions about Kismet might focus on the time spent training the robot, the training process actually works both ways, since 'Kismet's apparent randomness attests to the robot's reliance on the performative capabilities of its very particular "human caregiver"' (Suchman, 2007, p. 246). This analysis emphasises that, while the video footage of Kismet from MIT's website provides evidence of this robot's interactions with a reasonably large number of people, there would still seem to be some issues with its ability to read and respond to anyone who comes into the lab. This may lend credence to Weber's critique, since people reluctant to speak in 'motherese' while treating the robot as an infant (from a traditional perspective) may experience difficulty interacting with Kismet. However, it also suggests, as Suchman notes, that an extended period of learning how best to interact can be an important part of refining one's communication with even very humanlike robots.

More recently Suchman considers how 'the figure of the Human' is 'enacted in the design of the humanoid robot' articulating her concern, that the 'discourses and imaginaries' relating to humanoid robots 'will retrench received conceptions both of humanness and of desirable robot potentialities, rather than challenge and hold open the space of possibilities' (Suchman, 2011, pp. 119 and 130). She wonders if it is possible to 'refigure our kinship with robots...in ways that go beyond narrow instrumentalism, while also resisting restagings of the model Human' (Suchman, 2011, p. 137). Supporting the potential for such refiguration has been a central concern of this book, and, although many of its ideas were formulated before her call, I have focused my analysis on communication theory and practice, which Suchman argues should be 'a crucial site for reconceptualisation' (2011, p. 135). In particular, I have been concerned with rethinking communication, 'not as a medium through which an exchange of messages takes place', but instead as the name for 'the ongoing, contingent, coproduction of a mutually intelligible sociomaterial world' (Suchman, 2011, p. 135).

Suchman briefly introduces the possibilities of non-humanoid robots towards the end of her paper, arguing that communication between humans and such machines is 'explicitly that of evocation and response between different, non-mirroring, dynamically interconnected forms of being' (Suchman, 2011, p. 137). This idea has also been central to my analysis thus far, and continues to be important in considering the final two examples of human-robot interactions to be discussed in this book.

Long-term interactions with nonautonomous robots

The robots I have analysed up to this point rely on people for their construction and programming, but broadly, once operational, they are free to explore and interact with their surroundings in whatever ways their form and programmed abilities allow. In contrast, EOD robots are directly under the control of human operators much of the time, currently operating autonomously only to right themselves or retrace their steps. Communicating with these non-humanoid robots, which often resemble miniature tanks, involves the direct transmission of instructions using a controller with a wired or wireless connection. The robots respond to a human's instructions by moving as directed, and may also share information from their sensors via human-readable interfaces, such as monitors that show infrared camera images.

However, in spite of their limited autonomy, EOD robots are often assigned names and genders, and particular robots become important to EOD teams even though many examples of that make and model of robot will be deployed at any one time (Garreau, 2007). Although they are controlled by a human, EOD robots nonetheless reveal individual quirks in their behaviours during training and when on active duty (Garreau, 2007). These individual differences are not intentionally programmed but rather emerge as the robot operates in what are often challenging physical environments. The small differences in behaviour are therefore a result of the specific details of each robot's physical construction and general wear and tear, as well as particular patterns of damage that they have suffered. These unintentional behaviours are presented to observers (including the robot's operator) alongside the other radio-controlled movements the robots make, and it is this mixture of controlled and uncontrolled action that is read by people working with the robot (Figure 7.1).

EOD robots do not speak, but the directional 'gaze' of a camera, gestures of an arm and whole body movements can be read as kinesic acts of nonverbal communication, whether they are produced in response to an operator's command, or unintentional and uncontrolled. These embodied communications are interpreted by people as 'external signs of orientation and involvement' and can thus be understood as 'small behaviors' in Goffman's terminology (1972, p. 1). Having been identified as individuals, the experiences that team members and particular robots share build up to form a history that promotes either frustration with a robot that continually malfunctions or respect for one that performs reliably and well (The_Real_Opie; Singer, 2010).

It seems that EOD robots might be a reasonable answer to Goffman's question: '[w]hat minimal model of the actor is needed if we are to wind him up, stick him in amongst his fellows, and have an orderly traffic of behavior emerge?' (1972, p. 3). They show the type of activity that Cooren

FIGURE 7.1   *Packbot with soldier*
Source: US Navy image in the public domain.

suggests is required of an agent, in spite of being broadly nonautonomous. However, the suggestion that EOD robots communicate, given their current level of autonomy, is certainly heavily reliant on accepting that human interpretations, and at times their ventriloquisations, of the behaviours of these robots are more than just observations, but can be considered important in the operation of the human-robot team.

The connection between soldiers and individual EOD robots is developed during processes of training and operating together in the field. The same is true of human-dog teams, also used for explosive ordnance disposal, and in these teams both parties learn the other's communication style and abilities, as well as how to complete specific tasks. However, for the human-robot teams currently deployed, the onus is on the human to recognise the robot's specific abilities and learn to operate it as effectively as possible given that it may, for example, turn more easily in one direction than another.

In addition to noticing the nonverbal behaviours of EOD robots, it is also important to acknowledge the effect of their sociocultural positioning as lifesavers (Roderick, 2010). With each successful mission people's sense of trust in the robot increases, and the bond with an individual robot becomes stronger. As might be expected, soldiers with a damaged EOD robot that has saved lives on many occasions would prefer to have that particular robot repaired and returned, rather than receiving a replacement (QinetiQ, 2009; Singer, 2010). In contrast, while repair technicians may develop the sense that EOD robots are somewhat 'alive' when a repaired robot begins to move again, their response to replacing a robot damaged beyond repair is far more pragmatic (Garreau, 2007; QinetiQ, 2009).

Through continual training and deployment into a number of dangerous situations, the bond between human and EOD robot can be such that the robot becomes highly anthropomorphised. Humans and these robots work together over periods of weeks or months, rather than just for a series of experiments as in the case of humans and AUR. Individual EOD robots that perform well become regarded as brave and courageous team members, valued for their service in the line of duty, awarded medals and given funerals (Garreau, 2007; Singer, 2010; Garber, 2013). This level of bonding is regarded by some people as potentially dangerous (Ackerman, 2013; Carpenter, 2013; Garber, 2013; Waldman, 2013). The fear is that soldiers might not deploy a favourite robot even if it is the best suited for a particular task. However, there is no evidence that this has occurred, and it is equally possible that just as men and dogs continue to be deployed into dangerous conditions even in the knowledge that they might not return, the same will apply to robots.

In terms of developing fluent communication between humans and EOD robots, I would argue that this high level of respect and trust in a robot's abilities may become of practical value, assuming that EOD robots become more autonomous and aware of their surroundings and the task at hand in the future. This is because, as was the case for AUR, the human may need to be able to recognise that they are wrong, when the robot is right. This type of situation cannot arise at present, since EOD robots are broadly nonautonomous. However, these robots already have the potential to access very different information about the surrounding environment than humans and could be designed to respond autonomously to particular sensor readings, for example, regarding chemical or heat levels, even before the information is relayed to a person. This autonomous response would be designed to improve the efficiency of the robot, and of the human-robot team of which it is a part, but in order to be effective the human will need to recognise the potential validity of the robot's understanding of the situation, as was the case with AUR's double take, to make an informed decision about whether to intervene or allow the robot to choose its own course.

It is clear that through the process of working together over time humans bond with particular EOD robots, and while this can be read as strange, misguided or even potentially dangerous, it does mean that the human has become attuned to a particular robot. This should allow the team to become more effective, if only because the operator will know how best to direct that particular robot to perform a task given its specific quirks. However, I am more interested in what the future holds for these robots, if they do become more autonomous, able to make decisions for themselves based on their own sense of the surrounding world. If this happens, as seems possible, teams containing EOD robots and humans will benefit from operating with similar levels of respect and trust as were seen in human relations with AUR.

Moving closer to home with robotic floor cleaners

Of course, simple autonomous robots designed to complete tasks in human environments do already exist, albeit operating in considerably safer settings than EOD robots. In terms of people's everyday lives at home, the robots with which they are possibly most likely to have had some contact are robotic floor cleaners. While ownership of such a machine is still rare in many countries and cultures, by the end of 2013, one of the foremost commercial manufacturers of robots for personal, professional and military use, iRobot, had sold over ten million domestic service robots worldwide (iRobot Website: Our History). Most of these domestic robots are designed to vacuum or mop the floor, although the company also produces pool- and gutter-cleaning models.

As is the case for EOD robots, popular news reports indicate that a number of owners give their robotic vacuum cleaners names, assign gender to them and describe the movements, sounds and actions of the robots in ways that confer human or animal traits onto them (Kahney, 2003). This type of anthropomorphic and/or zoomorphic response has also been noted in scholarly articles in robotics, although not all researchers agree that such responses are an important factor when introducing these robots into homes. In particular, while Ja Young Sung et al. (2010) found that people did form relationships with their robotic vacuum cleaners, Julia Fink et al. (2013) argue that their research shows that the social aspect of people's relationship to such robots has been overestimated. Clearly not all people experience a robotic vacuum cleaner as part of their social world. They don't all individualise the robot by giving it a name and/or gender, talk to it or interact with it in any way other than through its standard interface. These owners would rather the robot simply operated as an autonomous tool needing minimal input and care, such as emptying its dirt collector or cleaning its brushes, which are unavoidable requirements at present (Fink et al., 2013).

There may be social and cultural factors at work in the differences between these studies, one being based in America and one in Europe. Alternatively, the results may simply be a product of the relatively small number of people recruited to take part in the trials. It seems likely that different people will have different responses to such robots, whether shaped by cultural understandings or attributable to personal taste; however, what is important to me is the way in which forming some level of social bond with the robot may help it to operate more effectively in the home by supporting a richer and more continuous communicative relationship with its owners.

In 'The Day We Brought Our Robot Home', Lydia Pyne (2014) provides a detailed and thoughtful description of the introduction of a new iRobot Braava floor-cleaning robot into her home. In common with people described in the research of Sung et al. (2010), Pyne and her husband decide to name their 'male' iRobot Braava, Isaac, reading 'his' actions in humanlike terms and interpreting Isaac as 'a chipper and earnest worker' who prefers 'open areas' over the confined space 'under the bookshelf'. This robot is regarded as a 'family member', as opposed to 'a motorized mop', and therefore as 'a living thing', maybe 'a pet' or 'a small child' (Pyne, 2014).

The findings of Fink et al. demonstrate that this type of anthropomorphic and/or zoomorphic response is not experienced by all owners of robotic floor cleaners. However, Pyne's article would seem to identify the benefits of reading the robot in this way, since she and her husband are happy to work with Isaac in order to help keep their house clean. In particular, they develop collaborative working patterns, setting up their rooms for efficient cleaning by moving furniture and rugs out of the way (Pyne, 2014). They are also attuned to listening out for when the robot 'whines in frustration' because it is stuck and requires assistance (Pyne, 2014). While it is impossible to argue whether the robot understands or benefits from these actions directly, the owners do, since a cleaner home is almost certainly the result of making the decision to work cooperatively with such a robot.

In analysing why her and her husband's response to the Braava is so marked, Pyne notes the importance of the Braava's 'outward behavior', in particular the robot's apparent ability to 'choose' in which direction to move. As Pyne says, it does not matter that underlying Isaac's 'choices' is a human-designed algorithm; the robot's movements are nevertheless read as intentional. In the case of robots such as the Braava, it is certainly not their form that makes them seem either humanlike or animal-like, since these robots resemble flattened boxes. Instead, it is the movements and sounds they make, both intentionally and unintentionally produced when stuck, that drive anthropomorphic and zoomorphic responses in some people. Pyne notes that she interprets Isaac as 'liking' to clean the floor and 'appreciating' when the job is made easier. While Pyne describes this level of anthropomorphism as important to making 'self-centric psyches' more comfortable by supporting their appraisal that the robot is content, I am more concerned to note that this type of response underlies useful understandings of the robot's communication and activity.

Pyne's recognition of Isaac's communicative acts, both intentional and unintentional, and her response, which is to attend to the robot's needs and help it if required, supports the success of this human-robot cleaning team. The story of Isaac therefore highlights the possibilities of collaborative teams between humans and relatively simple autonomous robots, within which the human recognises the robot's specific abilities and attends to how they can best help the robot to complete its task effectively. Again, as for EOD robots, the overarching control of the situation is with the human, but in this case the Braava is able to add its specific, and greatly valued, autonomous cleaning skills to the team.

In his discussion of activity and agents, Cooren notes that the decision to consider nonhuman agency 'is often criticized for questioning our tendency to firmly attach responsibility to action'; however, he argues that 'recognizing all forms of agency precisely allows us to speak of ethics and responsibility *in a very practical and incarnated way*' (2010, p. 6). All of the agents that actively take part in a given situation are therefore important in determining responsibility for what occurs. Although Cooren mentions ethics in this argument, and ethical questions do indeed come into play when thinking about the operations of EOD and other military robots in particular, the main focus of this book is on responsibility. For robots such as AUR and the Braava, part of what drives the interaction and ensures its success is the decision on the part of humans to take shared responsibility for the task at hand. For humans interacting with these robots, developing a respect for the robot's specific abilities and trusting it to do its job, while also playing one's own part to complete the task, not only supports fluency in interaction, but also ensures that the task is completed to the best of the team's ability.

Asymmetry, responsibility and reciprocity

Examples where humans and robots collaborate stress the importance of flowing communication alongside meaningful interruptions, this being reliant on the system of interaction itself, including the overarching context which may include a defined joint task, as well as the activities of individual agents. Considering communication from the perspective of both system and individual offers a way to understand how questions relating to responsibility, reciprocity and asymmetry in such relations are negotiated when performing a joint task. One of the vital things to recognise is that when humans and nonhumans form teams, the success of the team relies on the coordination of two very different skill sets. In all of the cases discussed in this book, whether they involve team-working between humans and dogs, or humans and robots such as AUR, EOD robots and the Braava, the human takes an overarching responsibility for how to tackle the task and is often the team member that issues ongoing instructions when required.

The communications between humans and animals, and between humans and robots, discussed in this book therefore draw attention to the asymmetry in the relation between self and other that arises because of their *'specific difference'* from each other (Haraway, 2003, p. 3). This asymmetry may result from the way in which '[e]ach participant in a communication situation is distinguished by a particular history and social position' (Young, 1997, p. 39). In interspecies relations, and in relations between humans and robots, it can also be linked with the disparate abilities of the agents. Paul Patton points out that, in order to be taken seriously, human-animal relations 'cannot be regarded as incomplete versions of human-human relations'; instead, they 'must be regarded as complete versions of relations between different kinds of animals' (Patton, 2003, p. 97). Following on from this, I would therefore argue that human-robot relations should be regarded as complete versions of relations between humans and machines.

Although humans are the ones who take responsibility for much of the time, it should be noted that the unequal endowment of power over the overall situation does not mean that dogs and robots have little to offer humans with whom they are working in teams. Hearne's experience with Gunner indicates that in this case the dog forced the human to take notice of the old scent trail he had discovered, and AUR's double take is sufficient for a human who has worked with the robot for a period of time to take notice and correct his mistake. These negotiations of asymmetry, within which the dog or robot interrupts and asserts themselves, are therefore vitally important, even as they sit within a situation of flowing communication for which the human has ultimate responsibility.

Other than at these specific moments, it is the human's controlling position that is most noticeable. In the case of robots, the level of control is made very clear as it is played out in the way that the robots are switched on or off by humans, although in the case of Fish and Bird they have to be caught first (a process that involves herding them into a corner such that the switch under the wheelchair seat can be operated). Even Data, who in general could be regarded as considerably more powerful than a human in terms of physical strength and some aspects of intellect, has an off switch, although its position is known only by a few members of the Enterprise crew. However, if the communication that develops through the dynamic interaction between human and dog, or human and robot, is responsive and respectful, as Hearne's assessment of Gunner and the positive result of AUR's double take demonstrate, the trust that the self develops for the other's difference and ability enables that other to take a powerful role of its own, to correct the course of the team's progress such that a joint task is completed successfully.

The longer-term relationships that have the potential to develop between humans and EOD robots or household robots such as the Braava emphasise the value of paying attention to the communicative acts of individuals, whether human or robot, as well as to communication that emerges within the system formed as human and robot interact in the context of a particular task and setting. Taking this dual perspective on communication has the benefit not only of explaining how non-humanoid robots can interact with humans to develop collaborative teams, but also suggests ways in which the asymmetries in those relations are flexible and allow the team to capitalise on human and robot skills and capabilities as appropriate to the task at hand.

Conclusion

Abstract: *The short conclusion to this book explains the basis for its somewhat eclectic analysis, which uses a range of traditions of communication theory, as well as considering the overarching conceptions of discrete state and dynamic systems methodologies.*

Sandry, Eleanor. *Robots and Communication*. Basingstoke: Palgrave Macmillan, 2015. DOI: 10.1057/9781137468376.0015.

This book has drawn out the ways in which different theories offer a variety of perspectives on communication. In particular, considering communication and the presence of difference between humans and robots emphasises the importance of holding in mind perspectives that focus on the system of interaction, alongside those that focus on the individual. The analyses of human-robot interactions with a range of communication theories presented here have purposefully made no attempt to accept one tradition over another as providing a better explanation of the communication taking place. Instead, the argument flows from the idea that any moment of communication can be fruitfully analysed from more than one perspective, usefully categorised by Craig (1999) as cybernetic, semiotic, critical, sociopsychological, sociocultural and phenomenological. However, as can be seen in the development of the argument in this book, it is also helpful to frame analyses using overarching conceptions, such as those that describe communication in terms of discrete states or continuous systems. This book therefore adopts a similar perspective to pluralism in communication theory and practice that Mary Midgley suggests is of value in science.

Midgley compares the proliferation of scientific theories with the maps that appear in the first few pages of a world atlas. Each map can be understood to show a different type of information about the world, for example, population or climate. She notes that 'if we want to understand how this bewildering range of maps works, we do not need to pick on one of them as "fundamental"' (Midgley, 2002, p. 82). Instead, Midgley suggests that a much better strategy is to consider 'why all these various maps are needed' and 'why they are not just contradicting one another' (2002, p. 82). One of the most important things to recognise is that 'the different maps' are actually engaged in 'answering different kinds of question, questions which arise from different angles in different contexts' (Midgley, 2002, p. 82). If this perspective is applied to the traditions of communication theory identified by Craig (1999), they can be seen to attend to many different questions, all of which may be asked about any particular moment of communication.

My overall concern has been to consider the possibilities of human-robot interactions, with a focus on articulating ways to regard otherness and difference as valuable within communication. The book's positioning of otherness in relations has relied upon the adoption of a broad conception of communication within which language, paralanguage and kinesics all play important roles. This appraisal of communication supports the acceptance of non-humanoid robots as expressive in ways that allow them to reveal their alterity in encounters understood through Levinas' conception of 'the face to face'. In addition, my argument has employed ideas about the agency that can be attributed to any participant that is active in a situation.

Being open to considering more than one communication tradition at once has provided valuable ways to gain new insight into interactions between humans and robots, whether the robots in question are humanoid or overtly non-humanoid in form. Within the interactions discussed in this book, self and other often meet in an asymmetrical relationship, this asymmetry being overtly present between humans and non-humanoid robots by virtue of their different forms, as well as the particular expressive communication codes that they use to reveal their personalities. However, adopting a dynamic framework for communication suggests that asymmetry can become a constantly changing aspect of the development of fluid relationships. In particular, the interruptions of nonhuman others have been shown to be valuable not only because they offer new insights into the other while also reinforcing the other's alterity, but also because such interruptions may draw attention to new ways to see particular tasks, or the world more generally. Although the exploration of communication theory in this book has been focused on exploring human-robot interactions, the approach to theory it has used may be productive when analysing communication processes and systems more broadly, as they occur between humans, animals and machines in any combination.

Bibliography

Ackerman, E. (2013) 'Soldiers can get emotionally attached to robots, and that may not be a good thing', *Spectrum IEEE*, 19 September.
ALAVs Website (no date) Available at: http://www.alavs.com/ (Accessed: 14 December 2014).
Asimov, I. (1977/1990) *Robot visions*. New York: ROC.
Asimov, I. (1953/1995) *The complete robot*. London: Voyager.
AUR Video (no date) Available at: http://techtv.mit.edu/collections/tonot/videos/497-anticipatory-perceptual-simulation-for-a-robotic-teammate (Accessed: 14 December 2014).
AUR Website (no date) Available at: http://alumni.media.mit.edu/~guy/aur/index.php (Accessed: 14 December 2014).
Bekoff, M. (2007) *The emotional lives of animals: a leading scientist explores animal joy, sorrow, and empathy—and why they matter*. Novato, Calif. : [s.l.]: New World Library; Distributed by Pub. Group West.
Benedetti, J. (1998) *Stanislavski and the actor*. New York: Routledge/Theatre Arts Books.
Blanchot, M. (1993) *The infinite conversation*. Minneapolis: University of Minnesota Press.
Boston Dynamics Website (no date) Available at: http://www.bostondynamics.com/robot_Atlas.html (Accessed: 14 December 2014).
Breazeal, C. L. (2002a) *Designing sociable robots*. Cambridge, Mass.: MIT Press.

Breazeal, C. L. (2002b) 'Regulation and entrainment in human-robot interaction', *International Journal of Experimental Robotics*, 21(10–11), pp. 883–902.

Breazeal, C. L. and Aryananda, L. (2002) 'Recognition of affective communicative intent in robot-directed speech', *Autonomous Robots*, 12(1), pp. 83–104.

Brooks, R. A. (2003) *Robot: the future of flesh and machines*. London: Penguin.

Brooks, R. A., Breazeal, C. L., Marjanovic, M., Scassellati, B. and Williamson, M. (1999) 'The Cog project: building a humanoid robot', in Nehaniv, C. (ed.) *Computation for metaphors, analogy, and agents*. Berlin; Heidelberg: Springer, pp. 52–87.

Capek, K. (1933) 'About the word robot', *Lidové noviny*, 24 December. Available at: http://capek.misto.cz/english/robot.html (Accessed: 16 January 2008).

Capek, K. (1920/2006) *R. U. R.* Adelaide: eBooks@Adelaide, University of Adelaide Library.

Carey, J. (1992) *Communication as culture: essays on media and society*. New York: Routledge.

Carpenter, J. (2013) *The quiet professional: an investigation of U.S. military Explosive Ordnance Disposal personnel interactions with everyday field robots*. PhD. University of Washington.

Chang, B. G. (1996) *Deconstructing communication: representation, subject, and economies of exchange*. Minneapolis: University of Minnesota Press.

Cherry, C. (1966) *On human communication: A review, a survey, and a criticism*. 2nd edn. Cambridge, Mass.: MIT Press.

Clark, D. (1997) 'On being "the last Kantian in Nazi Germany": dwelling with animals after Levinas', in Ham, J. and Senior, M. (eds) *Animal acts: configuring the humans in western history*. New York: Routledge, pp. 165–198.

Cooren, F. (2010) *Action and agency in dialogue passion, incarnation and ventriloquism*. Amsterdam; Philadelphia: John Benjamins. Available at: http://public.eblib.com/choice/publicfullrecord.aspx?p=623314 (Accessed: 10 December 2014).

Craig, R. T. (1999) 'Communication theory as a field', *Communication Theory*, 9(2), pp. 119–161.

DARPA Robotics Challenge Website (no date) Available at: http://www.theroboticschallenge.org/overview (Accessed: 14 December 2014).

Dautenhahn, K. (1998) 'The art of designing socially intelligent agents: science, fiction, and the human in the loop', *Applied Artificial Intelligence*, 12, pp. 573–617.

Davis, C. (1996) *Levinas: an introduction*. Cambridge, England: Polity Press.

Derrida, J. (2002) 'The animal that therefore I am (more to follow)', *Critical Inquiry*, 28(2), pp. 369–418.

Diehm, C. (2000) 'Facing nature: Levinas beyond the human', *Philosophy Today*, 44(1), pp. 51–59.

Disch, T. (1998) *The dreams our stuff is made of: how science fiction conquered the world*. New York: Free Press.

Eco, U. (1976) *A theory of semiotics*. Bloomington: Indiana University Press.

Fink, J., Bauwens, V., Kaplan, F. and Dillenbourg, P. (2013) 'Living with a vacuum cleaning robot: A 6-month ethnographic study', *International Journal of Social Robotics*, 5(3), pp. 389–408. doi: 10.1007/s12369-013-0190-2.

Finn, A. and Scheding, S. (2010) *Developments and challenges for autonomous unmanned vehicles: a compendium*. Berlin: Springer Verlag (Intelligent systems reference library, v. 3).

Fish-Bird Project Website (no date) Available at: http://www.csr.acfr.usyd.edu.au/projects/Fish-Bird/ (Accessed: 14 December 2014).

Flynn, C. P. (ed.) (2008) *Social creatures: a human and animal studies reader*. New York: Lantern Books.

Fogel, A. (1993) *Developing through relationships: origins of communication, self, and culture*. New York: Harvester Wheatsheaf.

Fogel, A. (2006) 'Dynamic systems research on interindividual communication: the transformation of meaning-making', *Journal of Developmental Processes*, 1, pp. 7–30.

Fong, T., Nourbakhsh, I. and Dautenhahn, K. (2003) 'A survey of socially interactive robots', *Robotics and Autonomous Systems*, 42, pp. 143–166.

Freud, S. (1919/2004) 'The uncanny', in Sandner, D. (ed.) *Fantastic literature: a critical reader*. Westport, Conn.: Praeger, pp. 74–101.

Garber, M. (2013) 'Funerals for fallen robots', *The Atlantic*, 20 September.

Garreau, J. (2007) 'Bots on the ground: In the field of battle (or even above it), robots are a soldier's best friend', *The Washington Post*, 6 May. Available at: http://www.washingtonpost.com/wp-dyn/content/

article/2007/05/05/AR2007050501009.html (Accessed: 13 March 2011).

Goffman, E. (1972) *Interaction ritual: essays on face-to-face behaviour*. London: Allen Lane.

Habermas, J. (1987) *Theory of communicative action*. Boston: Beacon Press.

Hanson, D. (2006) 'Exploring the aesthetic range for humanoid robots', in *Proceedings of the ICCS/ CogSci-2006 long symposium: towards social mechanisms of android science*. Vancouver, B.C., Canada, pp. 16–20. Available at: http://www.androidscience.com/proceedings2006/6Hanson2006ExploringTheAesthetic.pdf (Accessed: 19 October 2008).

Hanson Robotics Website (no date) Available at: http://hanson.robotics.com/ (Accessed: 18 October 2008, but no longer available).

Haraway, D. (1985/2000) 'A cyborg manifesto: science, technology and socialist-feminism in the late twentieth century', in Bell, D. and Kennedy, B. M. (eds) *The cybercultures reader*. London; New York: Routledge, pp. 291–324.

Haraway, D. (1988) 'Situated knowledges: the science question in feminism and the privilege of partial perspective', *Feminist Studies*, 14(3), pp. 575–599.

Haraway, D. (2003) *The companion species manifesto: dogs, people, and significant otherness*. Chicago: Prickly Paradigm Press.

Haraway, D. (2004) 'Cyborgs to companion species: reconfiguring kinship in technoscience', in *The Haraway reader*. New York: Routledge, pp. 295–320.

Haraway, D. (2008a) 'Encounters with companion species: entangling dogs, baboons, philosophers, and biologists', *Configurations*, 14(1), pp. 97–114. doi: 10.1353/con.0.0002.

Haraway, D. (2008b) *When species meet*. Minneapolis; London: University of Minnesota Press.

Hayles, N. K. (1999) *How we became posthuman: virtual bodies in cybernetics*. Chicago: University of Chicago Press.

Hayles, N. K. (2005) *My mother was a computer: digital subjects and literary texts*. Chicago: University of Chicago Press.

Hearne, V. (2000) *Adam's task: calling animals by name*. New York: Akadine Press.

Heyn, E. T. (1904) 'Berlin's wonderful horse', *The New York Times*, 4 September.

Hoffman, G. (2007) *Ensemble: fluency and embodiment for robots acting with humans*. PhD. Massachusetts Institute of Technology.

Hoffman, G. (2008) 'Achieving fluency through perceptual-symbol practice in human-robot collaboration', in *Proceedings of the 3rd ACM/IEEE International Conference on Human Robot Interaction*. Amsterdam, The Netherlands: ACM, pp. 1–8.

Hoffman, G. and Breazeal, C. (2010) 'Effects of anticipatory perceptual simulation on practiced human-robot tasks', *Autonomous Robots*, 28(4), pp. 403–423. doi: 10.1007/s10514-009-9166-3.

Honda (2007) 'ASIMO technical information'. Honda. Available at: http://asimo.honda.com/downloads/pdf/asimo-technical-information.pdf (Accessed: 1 September 2013).

Hornyak, T. (2009) 'Humanoid bot greets guests at Tokyo store', *CNET*, 20 October. Available at: http://www.cnet.com/au/news/humanoid-bot-greets-guests-at-tokyo-store/ (Accessed: 13 December 2014).

Hornyak, T. (2013) 'Asimo struggles on first day as science museum guide', *CNET*, 6 July. Available at: http://www.cnet.com/au/news/asimo-struggles-on-first-day-as-science-museum-guide/ (Accessed: 13 December 2014).

iRobot Website: Our History (no date) Available at: http://www.irobot.com/About-iRobot/Company-Information/History.aspx (Accessed: 14 December 2014).

Johnson, M. J., Feltovich, P. and Bradshaw, J. M. (2008) 'R2 where are you? Designing robots for collaboration with humans', in *Workshop on social interaction with intelligent indoor robots*. Pasadena, CA: ICRA, pp. 7–12.

Kahney, L. (2003) 'The new pet craze: robovacs', *Wired*. Available at: http://archive.wired.com/science/discoveries/news/2003/06/59249?currentPage=all (Accessed: 14 December 2014).

Kismet Videos (no date) Available at: http://www.ai.mit.edu/projects/sociable/videos.html (Accessed: 13 December 2014).

Knapp, M. L. and Hall, J. A. (2010) *Nonverbal communication in human interaction*. Boston, MA: Wadsworth, Cengage Learning.

Kuczaj, S. A., Ramos, J. A. and Paulos, R. L. (2002) 'Dancing on thin ice', *Behavioral and Brain Sciences*, 25, pp. 629–630.

Lakoff, G. (1980) *Metaphors we live by*. Chicago: University of Chicago Press.

Lasswell, H. D. (1948) 'The structure and function of communication in society', in *The communication of ideas: a series of addresses*. New York:

Institute for Religious and Social Studies and Harper & Brothers, pp. 37–52.

Leite, I., Martinho, C. and Paiva, A. (2013) 'Social robots for long-term interaction: a survey', *International Journal of Social Robotics*, 5(2), pp. 291–308. doi: 10.1007/s12369-013-0178-y.

Levinas, E. (1969) *Totality and infinity*. Pittsburgh: Duquesne University Press.

Levinas, E. (1980) *Otherwise than being, or, beyond essence*. The Hague: M. Nijhoff.

Levinas, E. (1989a) 'Is ontology fundamental?', *Philosophy Today*, 33(2), pp. 121–129.

Levinas, E. (1989b) 'The other in Proust', in Hand, S. (ed.) *The Levinas reader*. Oxford: Blackwell, pp. 160–165.

Levinas, E. (1990) *Difficult freedom*. London: Athlone Press.

Levy, D. N. L. (2007) *Love + sex with robots: the evolution of human-robot relations*. New York: HarperCollins.

Lingis, A. (1994) *The community of those who have nothing in common*. Bloomington: Indiana University Press.

Littlejohn, S. and Foss, K. A. (2011) *Theories of human communication*. 10th ed. Long Grove, Ill.: Waveland Press.

Maltby, R. (2003) *Hollywood cinema*. 2nd ed. Malden, Mass.: Blackwell.

Menzel, P. and D'Aluiso, F. (2000) *Robo sapiens: evolution of a new species*. Cambridge, Mass.: MIT Press.

Midgley, M. (2002) *Science and poetry*. London; New York: Routledge.

Moore, S. (1960) *An actor's training: the Stanislavski method*. London: Victor Gollanz.

Moore, S. (1968) *Training an actor: the Stanislavski system in class*. New York: Viking Press.

Mori, M. (1970) 'The uncanny valley', *Energy*, 7(4), pp. 33–35.

Nöth, W. (1990) *Handbook of semiotics*. Bloomington: Indiana University Press.

Patton, P. (2003) 'Language, power, and the training of horses', in Wolfe, C. (ed.) *Zoontologies: the question of the animal*. Minneapolis: University of Minnesota Press, pp. 83–99.

Peters, J. D. (1999) *Speaking into the air: a history of the idea of communication*. Chicago; London: University of Chicago Press.

Philmus, R. (2005) *Visions and re-visions: (re)constructing science fiction*. Liverpool: Liverpool University Press.

Pinchevski, A. (2005) *By way of interruption: Levinas and the ethics of communication*. Pittsburgh, Penn.: Duquesne University Press.

Poyatos, F. (1983) *New perspectives in nonverbal communication studies in cultural anthropology, social psychology, linguistics, literature, and semiotics*. Oxford; New York: Pergamon Press.

Poyatos, F. (1997) 'The reality of multichannel verbal-nonverbal communication in simultaneous and consecutive interpretation', in Poyatos, F. (ed.) *Nonverbal communication and translation: new perspectives and challenges in literature, interpretation and the media*. Amsterdam; Philadelphia: J. Benjamins, pp. 249–282.

Pyne, L. (2014) 'The day we brought our robot home', *The Atlantic*, 30 September. Available at: http://www.theatlantic.com/technology/archive/2014/09/the-day-we-brought-our-robot-home/380891/ (Accessed: 16 November 2014).

QinetiQ (2009) *Fast, powerful and versatile, high payload robot technology. TALON*. Farnborough: QinetiQ.

Reddy, M. (1979) 'The conduit metaphor: a case of frame conflict in our language about language', in Ortony, A. (ed.) *Metaphor and thought*. Cambridge: Cambridge University Press, pp. 164–201.

Roderick, I. (2010) 'Considering the fetish value of EOD robots: How robots save lives and sell war', *International Journal of Cultural Studies*, 13(3), pp. 235–253. doi: 10.1177/1367877909359732.

Russell, J. A. (1994) 'Is there universal recognition of emotion from facial expressions? A review of the cross-cultural studies', *Psychological Bulletin*, 115(1), pp. 102–141. doi: 10.1037/0033-2909.115.1.102.

Shanker, S. G. and King, B. J. (2002) 'The emergence of a new paradigm in ape language research', *Behavioral and Brain Sciences*, 25, pp. 605–656.

Shannon, C. and Weaver, W. (1948) *The mathematical theory of communication*. Urbana: University of Illinois Press.

Singer, P. W. (2010) *Wired for war: the robotics revolution and conflict in the twenty-first century*. New York: Penguin Books.

Smith, A. R. (1997) 'The limits of communication: Lyotard and Levinas on otherness', in Huspek, M. and Radford, G. P. (eds) *Transgressing discourses: communication and the voice of the other*. Albany, N.Y.: State University of New York Press, pp. 329–351.

Smuts, B. (2001) 'Living with animals', in Coetzee, J. *The lives of animals*. Princeton, N.J.; Chichester: Princeton University Press, pp. 107–120.

Smuts, B. (2008a) 'Between species: science and subjectivity', *Configurations*, 14(1), pp. 115–126. doi: 10.1353/con.0.0004.
Smuts, B. (2008b) 'Embodied communication in non-human animals', in Fogel, A., King, B. J. and Shanker, S. G. (eds) *Human development in the twenty-first century: visionary ideas from systems scientists*. Cambridge: Cambridge University Press, pp. 136–146.
Suchman, L. (2007) *Human-machine reconfigurations: plans and situated actions*. 2nd ed. Cambridge; New York: Cambridge University Press.
Suchman, L. (2011) 'Subject objects', *Feminist Theory*, 12(2), pp. 119–145. doi: 10.1177/1464700111404205.
Sung, J., Grinter, R. E. and Christensen, H. I. (2010) 'Domestic robot ecology: an initial framework to unpack long-term acceptance of robots at home', *International Journal of Social Robotics*, 2(4), pp. 417–429. doi: 10.1007/s12369-010-0065-8.
The_Real_Opie (no date) *Soldiers are developing relationships with their battlefield robots, naming them, assigning genders, and even holding funerals when they are destroyed, Reddit /r/Military*. Available at: http://www.reddit.com/r/technology/comments/1mn6wo/soldiers_are_developing_relationships_with_their/ccaxxsn (Accessed: 18 December 2013).
Turing, A. M. (1950) 'Computing machinery and intelligence', *Mind: A Quarterly Review*, 59(236), pp. 433–460.
Turkle, S. (2005) *The second self: computers and the human spirit*. Cambridge, Mass.: MIT Press.
Turkle, S. (2011) *Alone together: sociable robots, digitized friends, and the reinvention of intimacy and solitude*. New York: Basic Books.
Turkle, S., Breazeal, C., Dasté, O. and Scassellati, B. (2006) 'First encounters with Kismet and Cog', in Messaris, P. (ed.) *Digital media: transformations in human communication*. New York: Peter Lang, pp. 303–330.
Velonaki, M. and Rye, D. (2010) 'Human-robot interaction in a media art environment', in *Workshop: what do collaborations with the arts have to say about HRI?*. Osaka. Available at: hri.willowgarage.com/workshops/HRI2010/downloads/Velonaki.pdf (Accessed: 25 June 2010).
Velonaki, M., Scheding, S., Rye, D. and Durrant-Whyte, H. (2008) 'Shared spaces: media art, computing, and robotics', *Computers in Entertainment*, 6(4), p. 1. doi: 10.1145/1461999.1462003.
Waldman, K. (2013) 'Are soldiers too emotionally attached to military robots?', *Slate*, 20 September. Available at: http://www.slate.com/

blogs/future_tense/2013/09/20/military_bots_inspire_strong_ emotional_connections_in_troops_is_that_bad.html (Accessed: 14 January 2014).

Weber, J. (2005) 'Helpless machines and true loving care givers: a feminist critique of recent trends in human-robot interaction', *Information, Communication and Ethics in Society*, 3, pp. 209–218.

Wiener, N. (1948) *Cybernetics*. New York: John Wiley.

Wilden, A. (1972) *System and structure: Essays in communication and exchange*. London: Tavistock.

Wright, T., Hughes, P., Ainley, A., Bernasconi, R. and Wood, D. (1988) 'The paradox of mortality: an interview with Emmanuel Levinas', in *The provocation of Levinas: rethinking the other*. London; New York: Routledge, pp. 168–180.

Yeoman, I. and Mars, M. (2012) 'Robots, men and sex tourism', *Futures*, 44(4), pp. 365–371. doi: 10.1016/j.futures.2011.11.004.

Young, I. M. (1997) 'Asymmetrical reciprocity: on moral respect, wonder, and enlarged thought', in *Intersecting voices: dilemmas of gender, political philosophy, and policy*. Princeton, N.J.: Princeton University Press, pp. 38–59.

Index

accessibility, 14
acting theory, 80, 82–3, 87
Adlerian psychology, 93
AIBO (robotic dog), 43–4
Akhmatova, Anna, 63
alterity, 8, 37, 50, 52, 61, 63, 66, 90, 96, 99, 115
Andrew (robot), 16–7
android, 18
animal communication, 8, 30–5, 37, 39, 41, 43, 69
animal-like, 2, 9, 31, 43–4, 47, 90, 94, 109
animality, 37
anthropomorphic robot, 25
anthropomorphism, 8, 25, 38, 41–4, 56–8, 66–7, 83–5, 91–2, 94–7, 106, 108–9
antirhetorical rhetoric, 7
ape language research, 38
art installation, 9, 46, 51, 61, 63
ASIMO, 15–6, 18
Asimov, Isaac, 14–7
Atlas (robot), 95
AUR, robotic desk lamp, 9, 74–87, 97–8, 101–2, 106–7, 110–2
autonomous, 3, 9, 46–7, 59, 63, 77, 90, 102, 107–8, 110
Autonomous Light Air Vessels (ALAVs), 9, 46–61

Babe, 32
baboon, 38

backstory, 8, 63–4, 68, 97
Berk, Jed, 47–9, 53, 59
Berlo, David, 25
bipedal movement, 15, 17
Bird (robot), 9, 62–70, 72, 75, 87, 97–8, 111
Blanchot, Maurice, 50–1, 60, 65
blimp-like, 2, 9, 46–7
body language, 34–5, 39, 54, 68, 86
body movement, 17, 72, 78, 80, 82
Braava (robot), 2, 109–12
Breazeal, Cynthia, 23–5, 27, 30, 43, 77, 85, 92–4, 102
Bristol Robotics Laboratory, 17
Brooks, Rodney, 30, 81

Capek, Karel, 3, 13–7, 21
care robots, 4
Centre for Social Robotics, 63–4
Clark, David, 37, 53
cleaners, robotic, 4, 101, 107–10
Clever Hans, 34–5, 40
co-regulation, 69, 82
coding language, 19–20
cognitive science, 81
cognitive theory, 19
commonalities, 2, 50
commonality, 2, 8, 12–3, 26, 28–30, 43, 50, 56
communication, 1–10, 12–44, 46–66, 68–80, 82–90, 92–6, 98–115

communicators, 2, 5–8, 13, 20, 28, 40, 50, 58, 67, 101–2
companion species, 9, 39–41, 74–6, 79, 82, 84, 96
companions, 4, 8, 31
conceptual blindness, 7, 29
Cooren, François, 98–9, 104, 110
Craig, Robert T, 4–7, 20–2, 25, 29, 70, 114
critical (communication theory), 5, 21–2, 28–9, 114
cybernetic (communication theory), 5, 19–20, 22, 25, 28, 56, 70, 114
cybernetic-semiotic (combined communication theory), 20–2, 25–6, 28–9, 51, 55–6, 69, 72, 78
cyborg, 91

dance metaphor, 38, 41, 69–70, 72, 75, 82
Data, Lieutenant Commander (robot), 18–22, 25–6,72, 92, 111
decoding, 19
Delsarte (acting system), 80, 82–3
Derrida, Jacques, 36–7, 53, 66–7
dialogue, 5, 7, 9, 25, 27, 33, 58, 60, 62, 65, 68–9, 71–2, 80, 93
discourse, 5, 21, 28–9, 60
discrete state systems, 10, 69–70, 83, 87, 113–4
Disney, Walt, 32, 83
diversity, 5
dynamic systems, 7–8, 38, 68–73, 75, 80, 82–3, 87, 101–2, 115

education, 4
Ekman, Paul, 26
embodied communication, 39–41, 54, 76, 82, 104
emotional involvement (third stage of engagement), 72
emotional life, 17–8
emotions, 13, 16–7, 19, 21, 25–6, 49, 56, 80, 98
encounters, 2, 7, 9, 46, 49–50, 52–3, 55, 59, 61, 63, 97, 115

environment, 13–4, 27, 30, 48, 59, 76, 81, 107
EOD (explosive ordinance disposal robots), 4, 101, 103–8, 110–2
expressiveness, 15–17, 25, 83–5

face to face, 51–3, 55, 59–61, 63, 67, 91, 115
face, 6, 16–9, 23–8, 32, 36–7, 51–5, 59–61, 63, 65–7, 78, 83–4, 91, 97, 115
facial expression, 17, 20, 22–3, 25–7, 32, 52, 55, 57, 80–1
feelings, 13, 16–9, 21–2, 49, 84, 86
film, 3, 18, 32–3, 82, 84
Finn, Anthony, 101
Fish–Bird project, 63–9
Fish (robot), 9, 62–70, 72, 75, 87, 97–8, 111
Flipper (the dolphin), 33
fluency mode, 78–81, 85–6, 97, 102
Fogel, Alan, 69–72, 79, 82
friendship, 15–7, 23–4, 32, 85
functionalism, 20

gaze, 13, 36–7, 66–7, 84, 104
gesture, 18, 35, 40, 54–5, 68, 70, 77–8, 80, 104
Goffman, Erving, 35, 40, 54, 90, 104
greeting ritual, 38
Gunkel, David, 53, 91

Hanson, David, 17, 23–4, 43
Haraway, Donna, 9, 30, 37–41, 43, 67, 75, 87, 90–1, 96–7, 111
Hayles, N. Katherine, 91
Hearne, Viki, 41–4, 75, 86, 90–1, 96, 111–2
hermeneutical rhetoric, 7
history, 8, 39, 65, 71, 83, 102, 104, 108, 111
Honda, 15–6
human body language, 35, 39
human-animal communication, 8, 30–1, 33, 35, 37, 39, 41, 43, 69
human-dog, 8–9, 31, 74, 76, 87, 106
human-machine boundaries, 91

human-pet relations, 44
human-robot interaction, 73, 94
human-robot sociability, 24
human-shaped, 3, 15
humanlike, 2–3, 6, 8, 12–28, 30, 32, 43, 47, 55, 57, 77, 92–3, 95, 103, 109
humanoid, 2, 4, 6–9, 12, 14–5, 17–24, 28–30, 32, 43–5, 50, 55–7, 63, 69, 73, 90, 92–7, 99, 101–4, 112, 115
humans, 2–4, 6–9, 12–38, 40–1, 43–4, 47, 49–58, 60–1, 63–6, 68–75, 77, 79–80, 82, 84, 86, 89–93, 95–9, 101, 103, 106–7, 110–2, 114–5
hypodermic needle theory, 25

inauthenticity, 28
infant development research, 38
infant-caregiver relationship, 27
information processing, 5, 20, 70
installation, 9, 46–51, 53–5, 57–9, 61, 63–9, 71–2, 97–8, 101
inter-species language, 39
interaction, 5, 25, 35–40, 44, 54, 58–60, 64–5, 67, 70, 73, 75, 77–80, 85, 91–4, 99, 101–2, 110, 112, 114
interactive, 4, 47, 60
interruption, 51, 59–61, 64–5, 85–7, 110, 115
interspecies team, 42
intersubjective, 5, 20–1
Ishiguro, Hiroshi, 17

joint task (human and robot), 73, 75, 84, 110, 112
Jules (robot), 17–8, 22–3, 25–6, 57, 92

kinesics, 54–6, 60–1, 63, 69–70, 84, 90, 104
Kismet (robot), 23–7, 57, 72, 77, 81, 92–4, 102–3

laboratories, 4, 17, 18, 22, 73, 77, 81–2, 101
language, 2, 5, 17, 19–22, 25, 32–5, 37–9, 41, 52, 54–5, 59, 61, 63, 68–70, 72, 80, 84–6, 90, 93, 114

Lasswell, Harold D, 25
Levinas, Emmanuel, 6, 9, 28–9, 36–7, 46, 49–53, 55–6, 59–63, 65, 71–2, 87, 91, 96–7, 115
Levinasian, 49–50, 52–3, 55, 57, 60, 63, 84, 96–7
lifesavers, 106
Luxo Jr, 83–4

Machine Question, The, 91
machine, 3, 7–8, 10, 13, 15, 18, 20–1, 23, 52–3, 66–7, 85, 89–94, 102, 108
magic bullet theory, 25
materialism, 20
meaning making, 71, 82
Menzel, Peter, 93–5
metadiscourse, 5
Midgley, Mary, 114
Mitter, Nikhil, 47, 49, 53, 59
mobile telehone, 47, 58–9, 61
Moore, Sonia, 81
Mori, Masahiro, 22–3

news media, 4
noise, 20–1, 26
non-humanoid, 2, 7–9, 14, 32, 43–5, 50, 55–7, 63, 73, 90, 95–7, 101, 103–4, 112, 115
nonhuman, 52, 99
nonverbal communication, 7–9, 22, 31, 35, 38–9, 46, 53–6, 58–61, 72, 74, 79–80, 82, 104, 106

other, 1–10, 19–20, 22, 29–30, 32–3, 35–44, 46, 49–53, 55–7, 59–61, 63, 66, 71–2, 81–2, 84, 87, 90–9, 101–2, 104, 106, 108, 110–2, 114–5
otherness, 2, 4–7, 9, 29, 32, 44, 46–7, 49, 51, 53, 55, 57, 59, 61, 87, 91, 94, 97, 114

pain, 13, 16
paralanguage, 54–6, 60, 70, 84, 90, 93, 114
Paro (robotic seal), 43–4
partial understandings, 8, 57
partner, 4, 60, 77, 81, 86

Index

Path of Engagement (POE), 23
Patton, Paul, 111
people-friendly, 15
personality, 15, 22, 26, 43, 49, 57, 63, 67, 83-4, 94, 97, 115
persuasion, 5, 7, 29
pet, 43-4, 94, 109
Peters, John Durham, 2, 5-6, 20, 28-9, 35, 54
phenomenological, 5-7, 49, 71, 114
Pinchevski, Amit, 5-7, 29, 50-1, 55-6, 59-60
Poyatos, Fernando, 54, 56, 70, 90
prisoners (of war), 36-7
problems, 2, 5, 15, 17, 19, 21, 33, 55, 81
process of engagement, 71, 82
prosodic cues, 25, 93
prosody, 93
public spaces, 4
Pyne, Lydia, 108-9

R. U. R. (*Rossum's Universal Robots*), 3, 13-4, 16-7
radio-controlled, 3, 104
rationalism, 20
reactive mode, 78-9, 80, 85
read, 8, 13, 26, 35, 40-1, 54-5, 57, 83, 87, 102-4, 107, 109
reduplication of self, 29
rehearsal, 77, 82, 101-2
rescuers, 4, 33, 95
respect, 6, 10, 19, 38, 85-7, 97, 100, 104, 107, 110
responsibility, 6, 10, 21, 42, 55, 86-7, 98, 100, 110-1
rhetorical (communication theory), 4-7
robot personality, 84
robotic desk lamp, 9, 74-7
robotic technology, 4, 9, 14, 17, 23, 43, 49, 63, 74-7, 95, 101, 107-9
roboticist, 13, 17, 20, 22, 23, 28, 30, 43, 47, 63, 92, 94
robotics, 4, 8, 12-4, 17-8, 28, 63-4, 78, 92, 95, 108
Rossum's robots, *see R.U.R.*
running, 15, 79, 87
Rye, David, 63, 65-8, 72

Said, The, 59-60, 65
Saya (robot), 18
Saying, The, 9, 59-60, 62, 64-5
Scheding, Steve, 63, 101
Schramm, Wilbur, 25
science fiction, 3, 4, 13-4, 28, 57
self-other, 9, 30, 46, 49-50, 63
self, 2, 6, 9, 29-30, 46, 49-53, 55, 60, 63, 96-7, 109, 111-2, 115
semiotic (communication theory), 5, 19-22, 25-6, 28-9, 55-6, 80, 82-3, 114
servant, 14-15
shared space, 4
slave, 14, 21
small behaviors, 35, 40-1, 44, 54, 90, 104
Smuts, Barbara, 36-41, 43, 67, 75-6, 82, 84, 90-1
sociability, 15, 24-5, 94
sociable robot, 23, 25, 94
socially intelligent, 23-4
sociocultural (communication theory), 5, 25, 27-9, 51, 56, 72, 106, 114
sociopsychological (communication theory), 5, 25-6, 29, 51, 56, 114
Stanislavski (acting system), 80-3
Star Trek: The Next Generation, 18-9, 21
strangeness, 2-3, 44, 50-1, 54
Suchman, Lucy, 102-3
systems and information science, 19

team, 4, 9-10, 30, 31, 32, 40-2, 44, 73-4, 76-8, 82, 85-7, 92, 97, 100, 102, 104, 106-7, 110-2
television, 4, 18, 33
telos, 21
tempered (with respect to anthropomorphism and zoomorphism), 8, 57-8, 92, 96
'The Bicentennial Man', 16-7
'The Friends We Make', 14-7
thinking, 5, 8, 10, 15, 17, 21, 25, 41, 91, 99-100, 110

tone of voice, 39
transmission metaphor, 70
transmission of information, 2, 78, 90
trust, 10, 38, 42, 74–5, 77, 79, 81, 83,
    85–7, 97, 100, 106–7, 110, 112
Turkle, Sherry, 17–8, 92,
    94–5
turn-taking, 25, 33, 60, 69, 72–3,
    78–80, 93, 101–2

uncanny valley, 22–3

Velonaki, Mari, 63–8, 71–2
ventriloquising, 98, 106
violence, 29–30

walking, 15
Weber, Jutta, 94, 102–3
Williams, Stefan, 63
written instructions, 13

zoomorphism, 8, 43–4, 56–8, 67, 83,
    84–5, 92, 96–7, 108–9

The manufacturer's authorised representative in the EU is Springer Nature Customer Service Centre GmbH, Europaplatz 3, 69115 Heidelberg, Germany. If you have any concerns regarding our products, please contact ProductSafety@springernature.com

Printed and bound by CPI Group (UK) Ltd, Croydon, CR0 4YY

23/03/2026

02076355-0016